D0811875

THE
UNIVERSITY OF WINNIPEG
PORTAGE & BALMORAL
WINNIPEG 2, MAN. CANADA
DISCARDED

Photosynthesis

Photosynthesis

Eugene I. Rabinowitch
State University of New York at Albany

Govindjee
University of Illinois at Urbana

JOHN WILEY & SONS, INC.

NEW YORK · LONDON · SYDNEY · TORONTO

To the Memory of Robert Emerson

Library of Congress Catalog Card Number: 75-77830

Cloth: SBN 471 70423 7 Paper: SBN 471 70424 5

Printed in the United States of America

Preface

Fifteen years ago, when one of us completed a three-volume, 2000-page treatise on "Photosynthesis and Related Processes," one man still could have the ambition to read, digest, and analyze all the chemical, physical, and physiological research relevant to this fundamental life process on earth.

This would hardly be possible now. Scientific knowledge, which doubles on the average, about every ten years, has been accumulating even faster in the field of photosynthesis. Instead of 2000 pages, an up-to-date presentation of photosynthesis in all its aspects now may well require 10,000, and call for more sophistication in fields from theoretical physics through colloidal chemistry to plant breeding, than both of us together can possibly claim.

The present book deals more modestly, only with the physicochemical mechanism of the primary process in photosynthesis, and the enzymatic mechanisms closely associated with it. No attempt is made to cover these topics exhaustively—to describe all the related experiments or to discuss all the suggested physical and chemical hypotheses. Instead, only the main lines are traced.

On the other hand, we have provided a broad introduction into the subject by discussion of the basic physical and chemical concepts relevant to the understanding of photosynthesis, such as energy, entropy, free energy, bond strength, and oxidation-reduction potentials. We deal with the role of photosynthesis in the chemical household of the earth, its total yield, the carbon and oxygen cycles in nature, the origin and evolution of life, and its different modes—photoautotrophic, chemoautotrophic, and heterotrophic. This should make this book suitable for introduction into the field of photosynthesis for students with varying backgrounds—from physics to plant physiology—and also, we hope, for interested general readers and for high school students with advanced interest in biology.

Photosynthesis is one of the most fascinating achievements of biological evolution on earth. Life implies decrease in entropy—that means integrating disorderly materials into highly ordered units called "organisms." It, therefore, requires external supply of free energy; and the only widely available reservoir of such energy on earth is radiant energy reaching the planet from the sun. An amazing achievement of evolution had been to equip organisms with an apparatus for converting light energy into chemical energy. Without this provision, life on earth would have remained restricted to the rare spots where chemical free energy is readily available (such as volcanic springs).

Evolutionary development of photosynthesis has been a remarkable case of life "lifting itself by the bootstraps," because photosynthesis as we now know it, requires complex organic pigments—such as the green pigment, chlorophyll—as well as a number of organic catalysts (enzymes) that are themselves products of life. At the present time, we do not know of the formation of such compounds by processes other than biosynthesis in living organisms. Pigments, however, could have been first formed by a "prebiological" mechanism; in fact, derivatives of porphyrin (the mother substance of chlorophyll) have been identified among the products of electric discharges through media containing ammonia and simple carbon compounds. (A reducing atmosphere, containing such compounds, is widely assumed to have existed in "prebiological" times.) It is, nevertheless, remarkable that early organisms not only have incorporated these accidentally formed products, but also have discovered that they can be used to synthesize organic matter in light. One can well imagine biological evolution stopping, on many cosmic bodies, at the stage of "chemosynthesis," that is, with organisms depending for their existence on chemically unstable environment. Furthermore, if it is true that all atmospheric oxygen on earth has been produced by photosynthesis—as is now widely assumed—prephotosynthetic life must have depended on some unstable chemical systems not containing free oxygen. (Now, such systems are quite rare on earth.)

The striking nature of the evolutionary achievement involved in the development of photosynthesis is shown by our incapacity to repeat this process with the same efficiency outside the living cell, despite the many sophisticated techniques now at our disposal.

One fundamental difficulty is that photons of visible light are much smaller than the quanta of free energy stored, in photosynthesis, in the reduction of one molecule of carbon dioxide and liberation of one molecule of oxygen. Contributions of several (probably, eight) quanta must, therefore, be combined, through parallel or successive processes, to achieve a single elementary process of photosynthesis:

$$(CO_2 + H_2O \rightarrow \{CH_2O\} + O_2),$$

and we do not know as yet how to combine, in the laboratory, with a good yield, several quanta of visible light to produce a single energy-rich chemical molecule (or more precisely, a high-energy combination of two molecules, such as (CH_2O) and O_2).

Photosynthesis consists of three main stages: (1) the removal of hydrogen atoms from water and liberation of oxygen molecules; (2) the transfer of hydrogen atoms from an intermediate compound in the first stage to an intermediate in the third stage: (3) the conversion of carbon dioxide into a carbohydrate, $(CH_2O)_n$. The third stage is the best understood of the three. Its basic mechanism was established by 1954 by means of tracer experiments with radioactive carbon, ^{14}C. During the last fifteen years, major advances have been made in the understanding of the second stage—the transfer of hydrogen atoms from a derivative of water to a derivative of CO_2. It has been established that two light reactions are involved in this transfer and that a sequence of enzymatic reactions, in which cytochromes play a role, occur between them. Stage 1, the production of oxygen molecules, remains the least understood of the three. The next ten or fifteen years should see this darkness dispelled.

But even with the general outline of the mechanism known, innumerable details remain to be clarified, alternative paths must be explored, and qualitatively plausible hypotheses must be converted into quantitatively confirmed mechanisms. Generations of biochemists, biophysicists, and biologists will have their life work cut out for them until photosynthesis becomes fully understood—and mastered—by man. Good hunting!

E. Rabinowitch
Govindjee

January, 1969

Acknowledgments

We thank Mrs. Marion Bedell for her patience in typing the several drafts of the manuscript, Stanley Jones for making drawings, and G. S. Singhal and G. Papageorgiou for valuable discussions during the preparation of this book; also Drs. T. Bisalputra, Marcia Brody, Melvin Calvin, C. S. French, Elizabeth Gantt, W. Menke, R. Park, and T. Tanada for supplying some figures, and permitting us to use them.

We also thank all those, in the laboratory and at home—particularly our wives, Anya and Rajni—whose tolerance has permitted us to bring the manuscript of this book to completion.

E. Rabinowitch
Govindjee

Contents

Chapter 1

Photosynthesis: The Power Plant and the Chemical Factory of Life

Literally, photosynthesis means "synthesis with the help of light." This covers a variety of processes in organic and inorganic chemistry. However, the term is usually applied to one reaction only—the synthesis of organic matter by plants in light—a process also called "carbon assimilation." This is the basic process of life (at least as we know it on earth). It creates living matter out of inert inorganic materials, replenishes the reservoir of oxygen in the atmosphere, and stores the energy of sunlight to support the life activities of organisms. Its discovery is a thrilling chapter in the history of science.

About 1648, a Dutchman, *van Helmont,* grew a willow tree in a bucket of soil and found that the amount of soil did not diminish significantly, although a big tree was formed. He guessed that the material of the tree must have come from *water* used to wet the soil. In a book published in 1727 (called *Statical Essays, Containing Vegetable Statics, or, an Account of Some Statical Experiments on the Sap in Vegetation*), the great English minister-naturalist, *Stephan Hales,* surmised that plants drew a part of their nutrition from the *air.* Both views ran contrary to the long-accepted, Aristotelian view that plants feed on "humus" of the *soil.* Stephan Hales also suggested that *sunlight* may play a role in "ennobling the principles of vegetables."

Hales' and van Helmont's insights were remarkable. But before the advent of modern chemistry, they had to remain guesses, not provable by reliable experiment or by reference to well-established general laws.

THE AGE OF PNEUMOCHEMISTRY

Until the end of the eighteenth century, the different kinds of matter definitely known to man were *solids* or *liquids.* It was surmised that *air* also was something material, and that there existed different kinds of air, some "good" and some "bad," some able to support life, and some noxious or deadly. But not knowing how to weigh, transfer, mix, or separate the different kinds of air, chemists were baffled by reactions in which gases were formed or consumed. In fact, this was one of the weaknesses that made them alchemists rather than chemists! Metals rust. How would one explain it, not knowing that rusting is caused by the addition of oxygen from the air to the metal? Alchemists thought, not unnaturally, that in becoming rusty, and thus losing their value, metals must *lose* something, and they called this something *phlogiston.* Rusting, burning, and all other processes we now call oxidations were caused, according to them, by loss of phlogiston.

According to a law, first announced by Michael Lomonosov in Russia in 1748 and later by Antoine Lavoisier in France in 1770, the weight of the products of a reaction must be equal to that of the reactants. When Lavoisier found that rust weighs *more* than the metal from which it was formed, some adherents to the phlogiston theory, loath to abandon it, suggested that phlogiston must have negative weight! However, at about the same time, between 1770 and 1785, chemists in different countries of Europe, *Priestley* and *Cavendish* in England, *Scheele* in Germany, and *Lavoisier* in France, devised methods to catch gases, to transfer them from one container into another, and to determine their chemical and physical properties. The age of pneumochemistry (from the Greek word for breath) opened.

The air was found to consist of two main gaseous components. One was chemically reactive and was consumed in burning and respiration.

It became known as oxygen, the oxide-generating gas. The other was chemically inert, and became known as nitrogen, the niter-generating gas. It was also called azote, (from Greek word *azoē*), meaning "not sustaining life." Water was found to be a combination of oxygen with still another gas, which was called hydrogen, the water-generating gas. So-called fixed air, the asphyxiating gas produced by respiration of animals, burning wood, and heating of chalk, proved to be a combination of oxygen with carbon. It is now called carbon dioxide. Other gases, such as chlorine, carbon monoxide, and methane (swamp gas) were soon discovered. With these discoveries, the law of conservation of matter could be verified and the puzzle of phlogiston solved. Phlogiston was simply "minus oxygen." Chemistry began its transformation from a qualitative into a quantitative science.

IMPROVEMENT OF AIR BY PLANTS AND THE ROLE OF LIGHT

Joseph Priestley (1733–1804) was a nonconformist English minister. In 1791, his house in Birmingham was sacked by a mob because of his alleged sympathies with the French Revolution; in 1794, he emigrated to Pennsylvania. Early in the era of pneumochemistry, Priestley was engaged in pioneering experiments with gases, later described in two volumes called *Experiments and Observations on Different Kinds of Air.* The first volume, published in 1776, contains the discovery of the *improvement of air by plants:*

I have been so happy as by accident to hit upon a method of restoring air which has been injured by the burning of candles and to have discovered at least one of the restoratives which Nature employs for this purpose. It is vegetable as well as to animal life, both plants and animals had affected it vegetation. One might have imagined that since common air is necessary to in the same manner; and I own that I had that expectation when I first put a sprig of mint into a glass jar standing inverted in a vessel of water; but when it had continued growing there for some months, I found that the air would neither extinguish a candle, nor was it at all inconvenient to a mouse which I put into it.

Finding that candles would burn very well in air in which plants had grown a long time . . . I thought it was possible that plants might also restore the air which had been injured by the burning of candles. Accordingly, on the 17th of August, 1771, I put a sprig of mint into a quantity of air in which a wax candle had burned out and found that on the 27th of the same month another candle burnt perfectly well in it.

Two years later, in 1773, a court physician to the Austrian Empress Maria Theresa, a Dutchman, *Jan Ingenhousz* (1730–1799) visited London. He heard Sir John Pringle, then President of the Royal Society, describe in a lecture Priestley's experiments on the improvement of air by plants. Ingenhousz was so impressed that on the "earliest occasion" (which offered itself only six years later) he rented a villa near London, and spent there three summer months performing "over 500" experiments on the effects of plants on air. By October of the same year, he had not only completed a most momentous series of observations, but had also published a book, *Experiments Upon Vegetables, Discovering Their Great Power of Purifying the Common Air in Sunshine and Injuring It in the Shade and at Night.* Ingenhousz believed that he had made such important discoveries that immediate publication was needed to prevent somebody else from depriving him of priority. It was a hectic period in science, in which discoveries, made possible by experimentation with gases, crowded each other. The following is a quotation from Ingenhousz's summary of his findings on the action of sunlight on plants:

I observed that plants not only have a faculty to correct bad air in six or ten days, by growing in it, as the experiments of Dr. Priestley indicate, but that they perform this important office in a complete manner in a few hours; that this wonderful operation is by no means owing to the *vegetation* of the plant, but to the *influence of the light* of the sun upon the plant. I found that plants have, moreover, the most surprising faculty of elaborating the air which they contain, and undoubtedly absorb continually from the common atmosphere, into real and fine dephlogisticated air; that they pour down continually a shower of this depurated air, which . . . contributes to render the atmosphere more fit for animal life; that this operation . . . begins only after the sun has for some time made his appearance above the horizon . . . ; that this operation of the plants is more or less brisk in proportion to the clearness of the day and the exposition of the plants; that plants shaded by high buildings, or growing under a dark shade of other plants, do not perform this office, but on the contrary, throw out an air hurtful to animals; . . . that this operation of plants diminishes towards the close of the day, and ceases entirely at sunset; that this office is not performed by the whole plant, but only by the leaves and the green stalks; that even the most poisonous plants perform this office in common with the mildest and most salutary; that the most part of leaves pour out the greatest quantity of the dephlogisticated air from their under surface; . . . that all plants contaminate the surrounding air by night; . . . and by day; that roots and fruits have the same deleterious quality; . . . that that all flowers render the surrounding air highly noxious, equally by night the sun by itself has no power to mend the air without the concurrence of plants.

Priestley had been the first to observe the "improvement of air" by plants, but he had attributed this improvement to the slow process of "vegetation" of plants; while Ingenhousz noticed that it was due to a rapid chemical reaction that sunlight caused to occur in green leaves and stalks. In light, as well as in darkness, plants respire and consume oxygen similarly to animals. But when illumination becomes sufficiently strong, *liberation* of oxygen exceeds its uptake. Ingenhousz exaggerated when he described the gases produced by plant respiration as highly noxious; but at that time, no clear distinction was made between truly *poisonous* gases, such as carbon monoxide, and *inert* gases, which do not support life, like nitrogen or carbon dioxide.

THE PARTICIPATION OF CARBON DIOXIDE AND WATER

Ingenhousz's concern with priority proved to be justified. A Swiss pastor, *Jean Senebier* (1742–1809) published, in 1782, in Geneva, a rambling three-volume treatise, *Mémoires physicochimiques sur l'influence de la lumière solaire pour modifier les êtres de trois règnes, surtout ceux du règne végétal*. In this treatise, he described observations similar to those of Ingenhousz. Senebier, however, noted an important additional fact: the air-restoring activity of plants depends on the presence of "fixed air" (that is, carbon dioxide). He wrote:

I do not agree that common air of the atmosphere can be changed, in the leaves of vegetables, depositing there its phlogistic component, and leaving them after this cleansing as dephlogisticated air.

He suggested, instead:

> that fixed air dissolved in water is the nourishment that the plants extract from the air which surrounds them and the source of pure air which they provide by the transformation to which they submit the fixed air.

A special chapter in Senebier's book dealt with the proof that:

> air liberated by plants exposed to the sun is the product of the transformation of fixed air by means of light.

One more reagent, directly involved in photosynthesis, remained to be discovered. It was the most common one, water. Its universal presence in all organisms long hid the fact that it plays an active role in photosynthesis. Priestley, Ingenhousz, and Senebier had made only qualitative, or very rough quantitative observations. Lomonosov's and Lavoisier's law of conservation of matter called for more precise tests. *Nicolas Théodore de Saussure* (1767–1845), another scholar from the learned city of Geneva, was among the careful experimentalists who provided solid support for this law. In 1804, he published a treatise, *Recherches chimiques sur la végétation.* In this book, he showed that the sum of the weights of organic matter produced by plants and of oxygen evolved by them, is considerably larger than the weight of fixed air (carbon dioxide) consumed by them. Since plants used in his experiments received no supplies except air and water, he rightly concluded that photosynthesis must also involve water as a reagent, besides carbon dioxide. (One may note that he thus confirmed the guess made two hundred years earlier by van Helmont, as Priestley and Ingenhousz had confirmed those of Stephen Hales.)

The chemical equation of photosynthesis could now be written:

$$\text{fixed air} + \text{water} \xrightarrow[\text{green plant}]{\text{light}} \text{vital air } (+ \text{ plant nourriture, see below}) \qquad (1.1)$$

The "fixed air" term and the "nourriture" term in this equation were due to Senebier, the water term to de Saussure, the involvement of light and of green plant matter (which we now call chlorophyll, from the Greek words *chloros* for green and *phyllon* for leaf), to Ingenhousz, and the "vital air" term to Priestley.

Ingenhousz, in a later work, called "An Essay on the Food of Plants and the Renovation of Soils," published in 1796, translated the description of the whole phenomenon from the phlogiston language into the language of the new chemistry, founded by Lavoisier:

$$CO_2 + H_2O \xrightarrow[\text{green plant}]{\text{light}} O_2 + \text{organic matter} \qquad (1.2)$$

Senebier's "fixed air" became carbon dioxide, CO_2, Priestley's "vital air" became oxygen, O_2, and "plant nourriture" became organic matter, that is, chemical compounds containing carbon, hydrogen, and oxygen, which form most of a living body.

The improvement of air by plants was thus recognized as "photosynthesis," or synthesis of organic matter in light. In addition to the change in terminology, the 1796 pamphlet by Ingenhousz was the first to describe clearly the role of photosynthesis in the nutrition of plants. In 1789, in the second volume of his "Experiments," Ingenhousz had ridiculed Senebier's suggestion that fixed air is taken up in photosynthesis and contributes, as suggested in Eq. 1.1, to the nutrition of plants. In 1796, he reversed himself and recognized this uptake as the *only* source of carbon contained in the organic matter of plants.

Photosynthesis is a remarkable example of great discovery to which several men, of different national origin (English, French, Swiss, and Dutch), and different background (two ministers, a physician and a professional chemist) have contributed. A bitter controversy over priority soon developed, particularly between the ambitious, worldly court physician, Jan Ingenhousz, a master of biting irony, and the plodding provincial pastor, Jean Senebier. This controversy was long kept alive by their biographers. Priestley, perhaps the greatest experimentalist of the three, also found his protagonists, who resented any credit given to either Ingenhousz or Senebier. In truth, each one of these men has made an invaluable contribution to the discovery, and there is fame enough to share among them.

ENERGY STORAGE IN PHOTOSYNTHESIS

One more name must be added to complete the history of the discovery of photosynthesis: that of a German doctor, *Julius Robert Mayer*

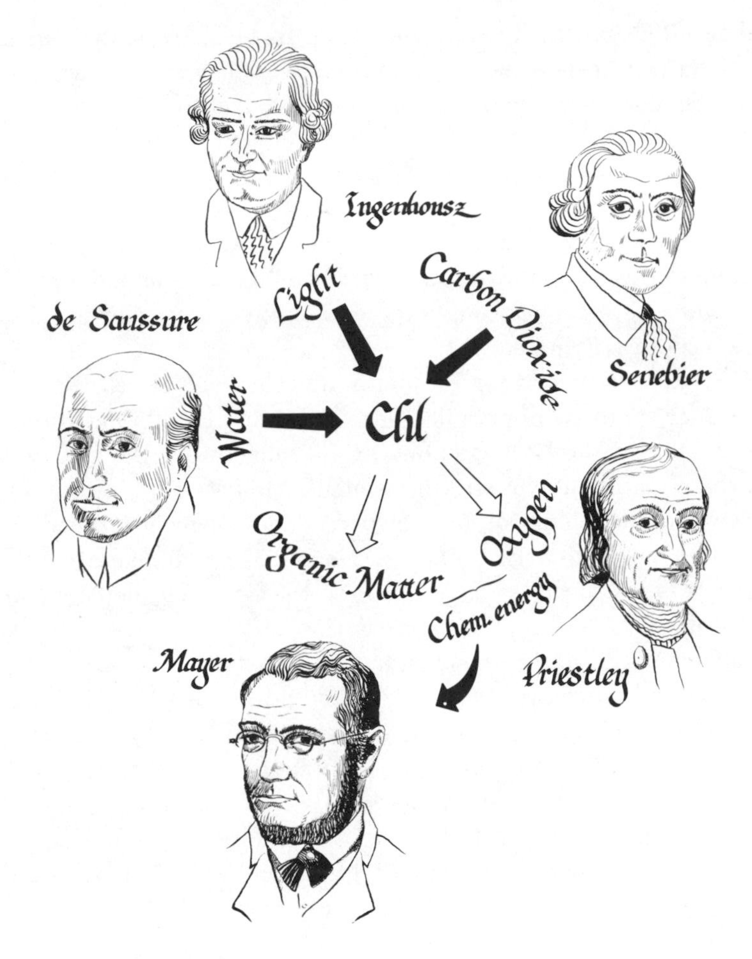

FIG. 1.1 The discoverers of photosynthesis.

(1814–1878), famous for his part in the formulation of the law of conservation of energy in 1842—sixty-six years after the discovery of photosynthesis (1776).

In carrying out photosynthesis, plants *store the energy of sunlight in the form of chemical energy.* Mayer saw in this conversion a particularly important illustration of the law of conservation of energy. In a pamphlet entitled, *The Organic Motion in its Relation to Metabolism,* published in 1845, he wrote:

Nature has put itself the problem how to catch in flight light streaming to the earth and to store the most elusive of all powers in rigid form. To achieve this aim, it has covered the crust of earth with organisms which in their life processes absorb the light of the sun and use this power to produce a continuously accumulating chemical difference.

These organisms are the plants; the plant kingdom forms a reservoir in which the fleeting sun rays are fixed and skillfully stored for future use; an economic provision to which the physical existence of mankind is inexorably bound.

The plants take in one form of power, light; and produce another power: chemical difference.

(Mayer used the term "power" where we would say energy, and "chemical difference" where we would say chemical energy.)

Before Mayer, only the chemical function of plants as *creators of organic matter* on earth, could be comprehended. After him, their physical function, that of *energy providers* for life, also became clear. The equation of photosynthesis could now be written as:

$$CO_2 + H_2O + \text{light} \xrightarrow{\text{green plant}} O_2 + \text{organic matter} + \text{chemical energy} \quad (1.3)$$

to represent not only the material balance, but also the energy balance of photosynthesis.

THE PRODUCTS OF PHOTOSYNTHESIS

Equations 1.1 to 1.3 are qualitative. They do not state the relative numbers of CO_2 molecules consumed and of O_2 molecules liberated by photosynthesis, and the composition of the organic matter produced is not specified. According to Avogadro's law (equal volumes of gases under the same pressure and temperature contain equal numbers of molecules) this question could be answered by measuring the ratio of the volume of CO_2 taken up ($-\Delta CO_2$) and the volume of O_2 liberated ($+\Delta O_2$). The first precise determinations of this ratio were carried out in 1864 by the French plant physiologist *T. B. Boussingault*. He worked with many different land plants, and found that the "photosynthetic ratios," $\Delta O_2/-\Delta CO_2$, were very close to unity for all of them.

As shown by Eq. 1.4, a photosynthetic quotient of 1 indicates that the organic matter produced by photosynthesis has the general composition of a carbohydrate, $C_n(H_2O)_m$:

$$n(CO_2) + mH_2O + \text{light} \xrightarrow[\text{plant}]{\text{green}} C_n(H_2O)_m + nO_2 + \text{chemical energy} \quad (1.4)$$

Equations similar to 1.4 can be written for the synthesis of organic products other than carbohydrates. However, those would require the volumes of oxygen evolved and of carbon dioxide consumed to be unequal. For example, fats are more strongly reduced (hydrogenated) than carbohydrates; consequently, in their formation the volume of oxygen released would be larger than that of carbon dioxide taken up.

For simple carbohydrates, such as glucose or fructose, $C_6H_{12}O_6$, $m = n$, so that Eq. 1.4 can be simplified to

$$CO_2 + H_2O + \text{light} \xrightarrow[\text{plant}]{\text{green}} (CH_2O) + O_2 + \text{chemical energy} \quad (1.5)$$

where (CH_2O) (= $\frac{1}{6}$ of $C_6H_{12}O_6$) signifies a unit of a carbohydrate molecule. That photosynthesis in green land plants leads to the synthesis of the carbohydrates, is illustrated by a simple experiment devised in 1864 by a leading German plant physiologist of the last century, *Julius Sachs*. He exposed one-half of a leaf attached to a plant to light and left the other in darkness. After some time, he placed the leaf in iodine vapor. The darkened half showed no change, but the illuminated half became dark-violet due to the formation of a starch-iodine complex. The American plant biochemist, *J. H. C. Smith*, demonstrated, in 1943, that in sunflower leaves, after one or two hours of photosynthesis, practically all the CO_2 taken up is found in newly synthesized carbohydrates, thus supporting Eq. 1.5 as representation of the overall process of photosynthesis.

In Chapter 17 we will see that compounds other than carbohydrates are also found among early products of photosynthesis. This must cause deviations of photosynthetic quotient from unity—which have been, in fact, observed. They are particularly strong in diatoms, which are known to store oil drops, as other plants store starch grains. It remains an open question whether the formation of fats (and of amino acids) occurs by side reactions competing with the completion of the main process (Eq. 1.5) or by rapid follow-up after its completion.

Chapter 2

Overall Chemistry of Photosynthesis; Autotrophic and the Heterotrophic Ways of Life

Photosynthesizing plants on earth are an immense organic-chemical factory and a giant energy transformer station. Let us estimate their overall annual performance.

TOTAL YIELD OF ORGANIC SYNTHESIS AND ENERGY STORAGE ON EARTH

Estimates of the total yield of photosynthesis on earth are by necessity only approximate. They start with measurements carried out, by a variety of methods, on limited areas of vegetation, for example, a grass plot, or a forest area, or a certain volume of water in a lake or ocean. Then, each result is multiplied by the total area or volume inhabited by this type of vegetation, and the products are added (see Table 2.1).

Unfortunately, the largest item is the one of which we know least. This is the production of organic matter in the ocean. The smaller figures

TABLE 2.1 Yield of Photosynthesis on Earth in Tons of Carbon Incorporated Annually into Organic Matter

Type of Vegetation	Area in Millions of Km^2	Annual Yield in Tons of C per Km^2	Total Yield in Billions (10^9) of Tons of C per year
Forests	44	250	11.0
Grassland	31	35	1.1
Farmland	27	150	4.0
Desert	47	5	0.2
Total on land	149[a]	—	16.3
Ocean	361	375[b] 62[c]	135[b] 22[c] 19[d]
Total	510	—	151[b] 38[c] 35[d]
Total, corrected for respiration[e]	—	—	173[b] 44[c] 40[d]

[a] Figures for land productivity in this table are from Schroeder, *Naturwissenschaften*, *7*, 8, 976 (1919). Since then, farm production must have increased considerably to provide for about doubled population!
[b] G. A. Riley, *J. Marine Research* (*Sears Found.*), *1*, 335 (1938); *2*, 145 (1939).
[c] A. Steemann-Nielsen, *Ann. Rev. Plant Physiol.*, *11*, 341 (1960).
[d] S. T. Pike and A. Spilhaus, *Marine Resources*, NAS/NRC Publ. 100 E (1962).
[e] A correction of +15% is applied to account for losses of carbon by respiration. This is a very uncertain figure (Steemann-Nielsen used +40% for the plankton).

given in Table 2.1 for the rate of oceanic synthesis have been derived from ^{14}C-tracer incorporation measurements, mainly by Steeman-Nielsen in Denmark; the higher ones, from measurements of oxygen production, mainly by Riley in the United States. A discrepancy of an order of magnitude exists between them. The newest estimates, included in Table 2.1, are "in between," but closer to Steeman-Nielsen's low figures.

In recent years, new interest in the ocean as a potential source of food for the growing world population has led to extensive oceanographic studies, covering physics, chemistry, and biology of the ocean. These studies have shown that the biological productivity of the waters in the ocean is far from uniform. According to a recent Russian review

(1966), based on oceanographic expeditions of Soviet ships, the photosynthetic production varies, in different parts of the ocean, by a factor of 16 (as compared to a factor of 50 on land, according to Table 2.1). The average value of oceanic photosynthesis is, however, closer to the lower limit than it is on land. (In other words, "deserts" include a greater proportion of the oceans than of the continents.)

The reason for wide differences in biological productivity of the oceans lies in variations in the supply of certain nutrient elements, such as nitrogen, phosphorus, iron, and manganese. This distribution depends strongly on vertical and horizontal water currents. Therefore, it is impossible to estimate reliably the total production of organic matter in the ocean from local measurements, and this may be the reason for the discrepancies in Table 2.1 Precise calculations must await the outcome of wide-range surveys, some of which are already under way, such as the international survey of the Indian Ocean. A committee of The International Biological Program (IBP) is planning a survey of all the world oceans.

As noted already by Ingenhousz, photosynthesis in plants is superimposed on the reverse process—respiration, that is, slow combustion of organic matter to water and carbon dioxide. A considerable part of the chemical energy stored by photosynthesis is converted in respiration into a special form—that of energy-rich phosphate. In this form, it can be used for various energy-consuming biological processes: production of mechanical work, chemical synthesis, osmotic work, and transmission of electric signals (Fig. 2.1).

An excess of photosynthesis over respiration is what permits growth of plants and storage of food reserves (starch, oil, fat) in them. In estimating the true yield of photosynthesis, the observed net yield must be corrected for plant respiration. With a crude corrrection of 15%, the total in Table 2.1 becomes 32–146 billion tons.[1]

These values of the total annual production of organic *matter* on

[1] The reader may ask: How can the amount of living matter on earth be approximately constant, if plant respiration takes care of less than ¼ of the products of photosynthesis? The answer is that *animal respiration* accounts for a significant part of the remaining ¾, particularly in the ocean, where animal life (zooplankton) is much more abundant than plant life (phytoplankton); the rest is destroyed by bacteria in the rotting of organic material on land and on the bottom of the sea.

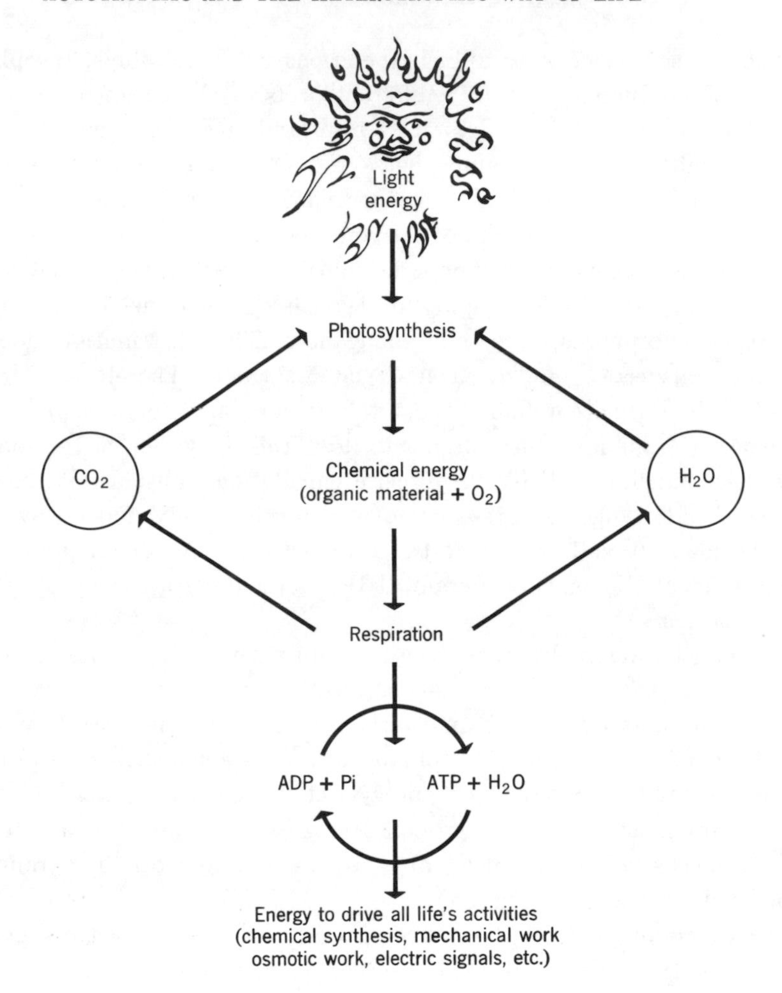

FIG. 2.1 Energy flow diagram showing conversion of light energy into chemical energy in photosynthesis, the formation of energy-rich phosphate in respiration and its use in energy-consuming biological processes. (ADP, adenosine diphosphate; ATP, adenosine triphosphate; Pi, inorganic phosphate; (P)OH has also been used for Pi, see Chapter 18.)

earth can be used to calculate the total *energy* storage by photosynthesis. A simple ratio exists between the chemical turnover and the storage of energy, determined by the amount of chemical energy stored in a unit mass of synthesized organic matter (not to forget the equivalent volume of oxygen!). Organic matter varies considerably in composition and energy content. However, its *average* composition is close to that of a carbohydrate $C_n(H_2O)_m$. Cellulose and starch are high-polymeric forms $(C_6H_{10}O_5)_n$, of the most common sugars, hexoses (obtained by elimination of one molecule of H_2O from each $C_{16}H_{12}O_6$ molecule), and much of the total weight of living organisms on earth is represented by cellulose, the structural material of trees. Carbohydrates are widespread also as food reserves in plants (starch) and in the liver of animals (glycogen). As an approximation, it is permissible to attribute to living matter on earth the average composition and average energy content of a carbohydrate.

All carbohydrates have approximately the same energy content of about 112 Kcal per gram atom (12g) of carbon contained in them. (One calorie is the amount of heat needed to heat one gram water by 1° centigrade; one kilocalorie is 1000 calories.)

Now, Eq. 1.5 can be rewritten by inserting the specific value of 112 Kcal for the term "chemical energy."

$$CO_2 + H_2O + \text{light} \xrightarrow[\text{plant}]{\text{green}} O_2 + (CH_2O) + 112 \text{ Kcal} \qquad (2.1)$$

where the brackets suggest that the equation refers to the formation of a (CH_2O) group in a carbohydrate molecule $(CH_2O)_n$.

Equation 2.1 indicates that storage of 112 Kcal is associated with the transfer of 12 grams (one gram-atom) of carbon into organic matter. This means a storage of about 9.6×10^6 Kcal in the formation of one ton (10^6 grams) of organic carbon. A total annual yield of 19 to 135×10^9 tons of organic carbon (Table 2.1) thus corresponds to storage of 1.8 to 13×10^{12} Kcal of chemical energy.

THE CARBON AND OXYGEN CYCLES ON EARTH

The total reserves of fossil coal on earth have been estimated as of the order of 10^{13} (ten trillion) tons. Photosynthesis adds annually something like one percent to the organic carbon now stored underground. Under today's climatic conditions, only a negligible part of this production is stored, while the overwhelming proportion is returned into the atmosphere and the ocean by respiration of plants and animals, by the burning of straw, wood, and dung, and by the activity of bacteria destroying dead plants and animals (Fig. 2.2).

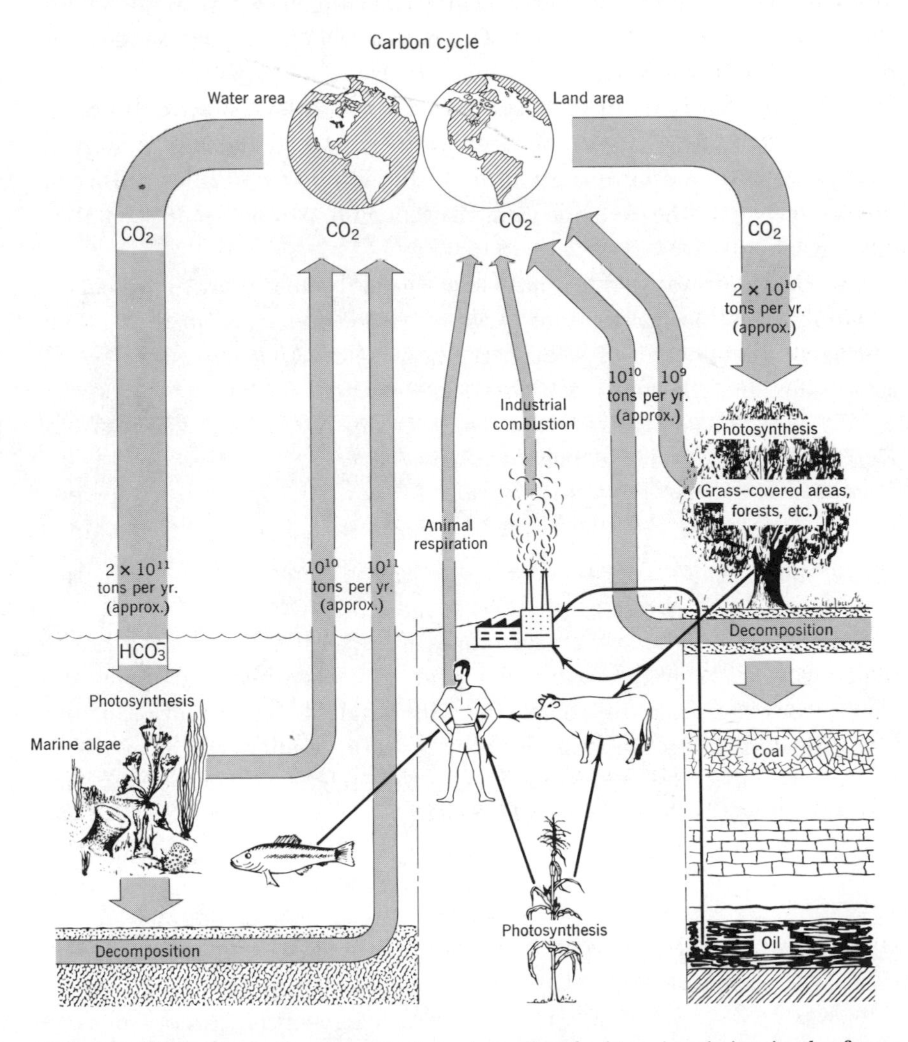

FIG. 2.2 The carbon dioxide cycle in nature. Zooplankton is missing in the figure between marine algae and fish. It accounts for a greater mass of marine life than the phytoplankton, and contributes largely to total respiration in the ocean. The figures for oceanic photosynthesis (and CO_2 liberation) used in this figure, were those of Riley, not the (almost ten times smaller) more recent figures of Steemann-Nielsen or Pike and Spilhaus. The CO_2 supply to the cereals feeding man and cattle is not shown.

An annual synthesis of about 5×10^{10} tons of organic carbon means liberation of about 13×10^{10} tons of oxygen into the air and fixation of about 20×10^{11} tons of carbon dioxide from the air and the oceans.[2] The amount of oxygen liberated is equivalent to about 0.05 percent of the atmospheric and oceanic store of free oxygen. Respiration and rotting of organic matter reverse this process. Together they create a more or less steady state distribution of carbon and oxygen between the free and the fixed state. Industrial combustion of fossil fuels makes a small, but not negligible addition to the natural cycle. At present, man makes annually about 3×10^9 tons of carbon dioxide by combustion. This is about one percent of the annual carbon dioxide consumption by photosynthesis (or of annual carbon dioxide liberation by respiration of plants and animals, and bacterial fermentation).

Because of photosynthesis and respiration, the elements carbon and oxygen go through closed natural cycles on earth (Fig. 2.3). Carbon passes by photosynthesis from gaseous or dissolved CO_2, into organic matter and thence by respiration back into the air or the ocean. Oxygen is consumed by the respiration of plants and animals and released again into air or water by photosynthesis. Since the total amount of oxygen in the air and in the ocean is ten times larger than that of carbon, the oxygen cycle turns around ten times slower than the carbon cycle. Every atom of available carbon on earth passes once every one (or a few) hundred years from the inorganic into the organic state. The same thing happens to every atom of oxygen present as O_2 in the air or in the water only about once every one or a few thousand years.

Actually, the situation is more complex. The carbon and oxygen cycles in the ocean revolve largely independently from the carbon and oxygen cycles on land, because the exchange of the gases CO_2 and O_2 between the air and ocean is slow.

There is, however, no doubt that all carbon in the organic matter now present on earth, and all oxygen now present in the air and in water, have passed several times through the organic cycle.

[2] Carbon dioxide is present in the oceans mainly as dissolved bicarbonate. It is considered half "free" and half "bound," because bicarbonate ion can give off one-half molecule of carbon dioxide:

$$HCO_3^- \rightarrow \tfrac{1}{2}CO_2 + \tfrac{1}{2}CO_3^{2-}$$

The other half becomes bound in the ion CO_3^{2-}, and usually made unavailable for photosynthesis by precipitation as an insoluble carbonate.

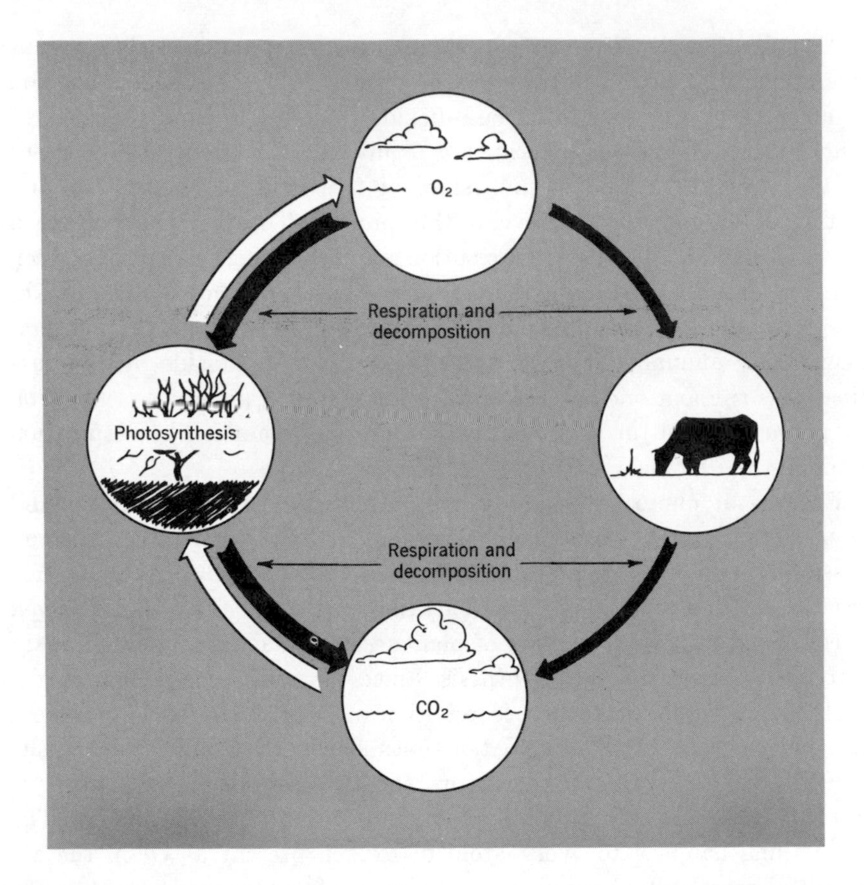

FIG. 2.3 The oxygen and carbon cycles on earth. (White arrows = photochemical reactions; black arrows = dark reactions.)

PHOTOAUTOTROPHIC, CHEMOAUTOTROPHIC, AND HETEROTROPHIC WAY OF LIFE

In certain locations on the earth's surface, there exist chemical energy sources of terrestrial origin, independent of sunlight. In regions of volcanic activity, hot water or hot gases escape from the earth. They often contain hydrogen sulfide, elementary sulfur, or hydrocarbons. Methane gas escapes from coal layers. (This is what creates the danger of mine explosions.) Free hydrogen is found in the soil, where it is formed by the activity of soil bacteria. The ultimate origin of these chemically unstable, combustible products (that is, products that contain free chemical energy when in contact with oxygen), may be either the heat of the interior of the earth (heat of volcanoes or hot springs), or the activity of now existing organisms, or the activity of organisms that have existed on earth in the past, and whose remains have been preserved underground in the form of coal, peat, or oil. Whatever its origin, this energy could be used—and in fact, is used—to support life in the form of so-called autotrophic, that is, self-nourishing bacteria. These bacteria are of two types: one kind merely substitutes a reducing chemical compound for the water used in photosynthesis of higher plants, in a process called *bacterial photosynthesis.* These organisms, which do not require *oxygen* but do require *light,* were discovered by the Russian microbiologist *S. Vinogradsky* in **1889**. *C. B. Van Niel,* in a series of articles published in **1930** and later, proposed the following generalized equation for their photosynthesis:

$$CO_2 + 2H_2A + \xrightarrow[\text{light}]{\text{bacteria}} (CH_2O) + 2A \text{ (or } A_2) + H_2O \qquad (2.2)$$

where A can be sulfur, or an organic radical. One particularly interesting case is "A equals nothing"—bacterial photosynthesis with molecular hydrogen as reductant. (Higher plant photosynthesis is a special case of Eq. **2.2** in which A is oxygen.)

The second type of autotrophic bacteria catalyze the oxidation of various compounds (usually by oxygen, but occasionally by nitrate, sulphate, or even carbonate) and utilize the energy of oxidation for the reduction of carbon dioxide to organic compounds. This process is called *chemosynthesis.* These organisms require *no light* for their growth, but they do require *oxygen* (or another oxidizing agent).

We call organisms capable of living by means of photosynthesis, *photoautotrophic* (or self-supporting with the help of light). Organisms capable of living by means of nonphotochemical energy—releasing chemical processes are called *chemoautotrophic,* whereas organisms able to live only by consuming other live or dead organisms, or organic food, are *heterotrophic.* All animals, including man, as well as the fungi, most bacteria, and all viruses, belong to this latter, fundamentally parasitic kind. They cannot synthesize their own food, but rely on autotrophic organisms to prepare it for them.

ORIGIN AND EVOLUTION OF LIFE

The present picture of the evolution of life on earth, originating with *J. Haldane* in England and *A. I. Oparin* in Russia, assumes that in the early history of the earth, when life was absent, the earth's atmosphere contained no oxygen and had reducing properties. Hydrogen, ammonia, hydrogen sulfide, and simple hydrocarbons abounded in this atmosphere. Under the influence of electric discharges (during thunderstorms) and of ultraviolet light from the sun, increasingly more complicated organic compounds, such as amino acids, peptides, and porphyrin derivatives were synthesized. The ocean became a primeval "broth" filled with a great variety of organic compounds. At a certain stage of this chemical evolution, complex molecules arose with a capacity of assimilating simpler molecules and arranging them into copies of themselves. This is still done now by one kind of molecules only—the nucleic acids. When the nucleic acid molecule of a virus is injected into a living cell, whose interior takes the place of the primeval soup, the virus picks up simpler building stones, nucleotides and amino acids, and builds up complete new viruses, thus exhausting and soon destroying the cell. Recently, such self-replication of nucleic acids was achieved also in vitro. The original living molecules must have similarly scavenged organic material in the primeval broth. Theirs was a heterotrophic way of life; it could not last forever, since the broth had to become thinner and thinner. Some kind of "mutation" must have interfered to prevent the end of all life; it converted heterotrophic into autotrophic organisms, able to build organic molecules from inorganic building stones. At first, perhaps, this energy-consuming process utilized the chemical energy of simple inorganic reactions—that is, the organisms were chemoautotrophs. Later, they found a way to utilize the energy of light, and photoautotrophs evolved.

It seems plausible that, at first, hydrogen for the reduction of CO_2 was derived from the then still abundantly available reducing gases, such as H_2S or H_2; that is, the earliest photoautotrophic organisms may have been *photosynthetic bacteria*. However, the availability of such reductants also decreased with time. Another mutation must have intervened, leading to a new kind of organism, the true plant, capable of using the practically unlimited supply of water as a source of hydrogen.

In this new era, the atmosphere began to fill up with oxygen, and a new type of heterotrophic existence became possible, based on aerobic respiration, rather than on simple ingestion of ready-made organic compounds. Animals and fungi were added to viruses in the world of heterotrophs, and plants themselves began to depend on respiration as source of biological energy.

Chapter 3

Overall Energetics of Photosynthesis

ENERGY, ENTROPY, FREE ENERGY

The basic concept of physics is *energy;* that means, capacity to produce work. The basic law of physics is the law of conservation of energy (as the basic law of chemistry is that of conservation of matter). This law asserts that *the amount of energy in the universe is constant.* This sweeping statement is a generalization of the more restricted *first law of thermodynamics,* which proclaims the constancy of the sum of mechanical and thermal energy in systems containing energy only in these two forms.

According to the law of conservation of energy, the latter cannot be created or lost. However, different *forms* of energy, such as electromagnetic, chemical, or nuclear, are interconvertible. All of them can be converted to mechanical energy and made to produce work; that is, to impart acceleration to material bodies.

However, from the point of view of mutual convertibility of the different forms of energy, one important distinction exists. All forms of energy are *completely* interconvertible, except one: thermal energy. This is easy to understand, if one considers the molecular structure of matter. Thermal energy is the mechanical energy of the chaotic motion of atoms and molecules. Increase in intensity of this motion is what we call temperature rise. The chaotic character of thermal agitation is the reason for the limited convertibility of thermal energy into other forms of energy. In a resting mass of gas or liquid, as many molecules move, at a given time, in one as in any other direction. The law of conservation of energy would not be violated if the thermal energy of a mass of water in the ocean would be converted into the kinetic energy of a ship riding this water. Water would cool down and the ship would speed forward. All that is needed for this is to give the chaotic motion of water molecules a preferred direction. We know from experience that this does not happen; and we never expect such conversion of disorder into order to occur spontaneously. This conviction is expressed in a fundamental principle that says, in essence, that order does not arise spontaneously out of disorder; to create order out of disorder, a certain amount of energy must be expended (that is, converted into heat). This is the *second law of thermodynamics;* it says that in a closed system (a system not exchanging energy with the surroundings) heat cannot be converted into work without some other change in the system involving an increase in disorder and balancing the increase in orderliness associated with the conversion of heat into work.

For precise formulation, this principle requires a quantitative measure of disorder. This measure is called *entropy.* (The measure of *order* is sometimes called *negentropy.*) The precise definition of entropy has to do with the fact that every orderly state of a system of many particles is in some way unique, while disorderly states can be realized in a multitude of different ways. For example, the state in which all girls sit on the right side and all boys on the left side of the classroom is more orderly than the state in which sexes are mixed at random. It is obvious that the first distribution can be arranged through fewer different individual seat assignments than the second. Entropy, S, is defined (apart from a proportionality constant, k) as the logarithm of the number (n) of different ways in which a particular state of an assembly of many particles can be achieved (Eq. 3.1).

$$S = k \ln n \tag{3.1}$$

(By taking the logarithm, ln, rather than the number itself, entropy becomes, like energy, an *additive,* rather than a *multiplicative* property, so that the entropy of two systems together is equal to the sum of the entropies of the same two systems taken separately.)

The thermodynamic state of a system of many particles is thus defined (a) by its content in energy (E), and (b) by its content in entropy (S). The higher the entropy of a system, the more disorderly it is, and the less of its thermal energy is available for conversion into usable forms of energy—mechanical, chemical, or electric. The "usable" part of total energy (E) of a system is called its *free* energy (F). Thermal energy (heat) is the only kind of energy that is only partly free. A change in free energy (ΔF) in a process is related to the change in entropy (ΔS) and the change in the total energy (ΔH, or enthalpy) by the following equation:

$$\Delta F = \Delta H - T\,\Delta S \tag{3.2}$$

where T is the so-called absolute temperature. Its definition and implications will be discussed below.

In a closed system, spontaneous processes cannot go in the direction of decreasing entropy (that is, increasing order) because this would contradict the second law of thermodynamics. After every spontaneous process in such a system, entropy must increase. This means that a less probable, or more orderly, state must go over into a more probable, less orderly state.

When we deal, as we usually do, with systems that are *not* closed (systems that can exchange energy with the surroundings), more complex formulations of the second law of thermodynamics become necessary. One important case is that of a reactive system enclosed in a reservoir of constant temperature and exchanging heat with it, such as an organism in air. We can imagine, for example, a sphere, *Sy*, inside which a combustion, or another chemical reaction can be produced (see Fig. 3.1). Let this sphere be submerged in a water reservoir, *R*, so large that its temperature is practically unaffected by taking up, or supplying, the enthalpy of the reaction, ΔH. The process is thus conducted *isothermally*, the same temperature prevailing before and after the reaction; but a certain amount of thermal energy, ΔH, is transferred from the system to the reservoir or vice versa. If thermal energy is lost by the system to the reservoir, ΔH is given the negative sign ($-\Delta H$); a positive sign ($+\Delta H$) indicates that energy is acquired by the system from the reservoir.

Since entropy is additive, the change in total entropy of the reservoir and the system can be divided into two terms: one is the change in

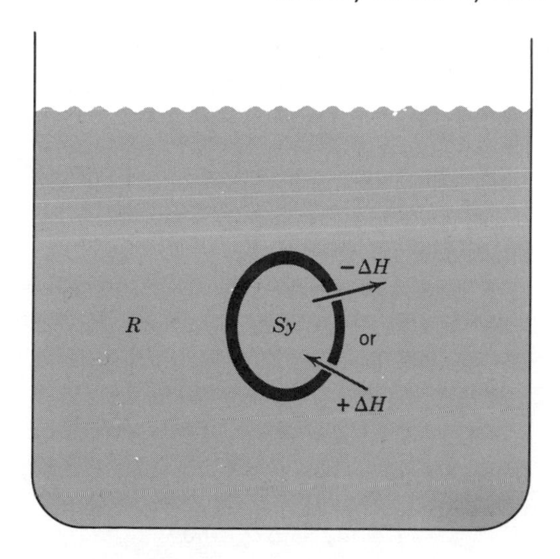

FIG. 3.1 Isothermal reaction. (*Sy*, system; *R*, reservoir.) Thermal energy, ΔH, is passed *from* the reacting system, *Sy*, to the reservoir, *R*, if the reaction is exothermal; from the reservoir to the system if it is endothermal. The total change in *entropy* (disorder), ΔS, is $\Delta S_{Sy} + \Delta S_R$. The second term is equal to $-\Delta H/T$. For reaction to proceed, the total entropy (disorder) must increase: $\Delta S_{Sy} - \Delta H/T \geq 0$; or, $\Delta H - T\Delta S_{Sy} \leq 0$; or, $\Delta F \leq 0$.

the entropy of the reservoir, ΔS_R, caused by the addition or subtraction of thermal energy, ΔH; it is positive for "exothermal" (heat-liberating) reactions, and negative for "endothermal" (heat-consuming) reactions. The second term is the change in entropy of the reacting system (ΔS_{Sy}). This term is positive if the reaction leads to a decrease of orderliness and negative if there is an increase in orderliness.

Since, together, the reacting system and the reservoir form a closed system, the condition that must be satisfied for the reaction to proceed is

$$\Delta S = \Delta S_{Sy} + \Delta S_R \geq 0 \tag{3.3}$$

The entropy increase in the reservoir, $+\Delta S_R$, caused by transfer of thermal energy, $-\Delta H$, from the system to a reservoir with a temperature T, or the entropy decrease in the reservoir, $-\Delta S_R$, associated with the transfer of energy $+\Delta H$ from the reservoir to the system, is proportional to ΔH (Eq. 3.4). Equation 3.4 provides the thermodynamic *definition* of absolute temperature T: the proportionality constant in this equation is the reciprocal of T.

$$\Delta S_R = \frac{1}{T}(-\Delta H); \quad \text{or,} \quad T\,\Delta S_R = -\Delta H \tag{3.4}$$

Equation 3.4 states that the lower the temperature of the reservoir, the greater is ΔS_R (change in entropy). In other words, the lower the temperature, the larger is the part of thermal energy of a reservoir not available for conversion into mechanical or other free energy. A given amount of thermal energy becomes less "useful" as the temperature decreases. We can rewrite Eq. 3.3 with the help of 3.4 as

$$\Delta S_{Sy} - \frac{\Delta H}{T} > 0,$$

or, multiplying by T,

$$T\,\Delta S_{Sy} - \Delta H > 0 \tag{3.5}$$

or

$$\Delta F \equiv (\Delta H - T\,\Delta S_{Sy}) < 0; \quad \text{or,} \quad -\Delta F = -\Delta H + T\,\Delta S_{Sy} > 0 \tag{3.6}$$

where ΔF is the change in *free* energy associated with the process (as in Eq. 3.2). The absolute value of $-\Delta F$ is greater than that of the $-\Delta H$ (enthalpy) of the process, if the entropy change, ΔS_{Sy}, is positive; ΔF is smaller than ΔH if ΔS_{Sy} is negative. We note that at $T = 0$ (absolute zero, $-273°C$), when all thermal motion stops, ΔF becomes equal to ΔH. Equation 3.5 is the form of the second law of thermodynamics most often applied to chemical reactions, becauase they are usually carried out without insulation from the atmosphere, the latter acting as a thermal reservoir. Equation 3.6 says that *in order for a reaction to proceed spontaneously, its free energy change, ΔF, must be negative.* ΔF is the chemical energy that can be converted into other useful forms of energy in the reaction. As noted above, the absolute value of $-\Delta F$ is greater than that of $-\Delta H$ if the reaction proceeds with a decrease in orderliness ($\Delta S_{Sy} > 0$, that is, the second term in Eq. 3.6 is positive) and smaller if the reaction proceeds with an increase in orderliness ($\Delta S_{Sy} < 0$, that is, the second term in Eq. 3.6 is negative). Thus for an isothermal reaction to proceed spontaneously, it must have either a sufficiently large *negative* ΔH (give off enough heat) or a sufficiently large positive ΔS (that is, it must lead to a sufficient increase in disorder) to make the sum $\Delta F = \Delta H - T\,\Delta S$ negative.

This explains why some endothermal reactions, which consume rather than evolve heat, nevertheless proceed spontaneously. For example, some salts dissolve in water with consumption of heat (so that the solution cools down, if it is not supplied with heat from a reservoir). In this case, the increase in *entropy* (increase in disorder), caused by breakdown of the crystalline salt structure into free ions, overbalances the acquisition of heat energy needed to maintain constant temperature.

In chemical reactions with *large* enthalpy (ΔH) changes, the entropy term, $T\,\Delta S$, becomes relatively insignificant. For example, in photosynthesis, the stored energy ΔH, is about 112 Kcal/mole (Eq. 2.1), while the stored free energy, ΔF, is about 120 Kcal/mole. (This figure applies to open air conditions, when the proportions of CO_2 and O_2 are 0.03% and 21% respectively; the concentration of $C_6H_{12}O_6$, which also affects the free energy, is arbitrarily taken as 1 mole/liter.) The ΔF of photosynthesis is slightly greater than ΔH (that is, ΔS is negative), because the system consisting of large sugar molecules and small oxygen molecules is somewhat more orderly than the system consisting only of the small molecules, H_2O and CO_2.

In order to increase the orderliness of a system, and thus to decrease its entropy, energy must be supplied from the outside. Inversely, useful energy can be extracted from a system if its entropy increases. When thermal energy is passed from a body of *higher temperature* to one of lower temperature, equalizing the average thermal energy of the molecules and thus destroying the orderliness inherent in the segregation of the faster from the slower molecules, some free energy can be derived from this passage. This is the principle on which the heat engines operate. The same is true of equalization of *pressure* or of *concentration* between two parts of a system. The amount of free energy available from various equalization processes depends on the degree of inequality in temperature, pressure, or concentration available in the system. This is why, in a heat engine, the amount of work that can be derived from the transfer of a certain amount of heat from a hot to a cold reservoir (the "efficiency" of the engine) is larger the greater the temperature difference between them.

When we say that life consumes energy, we really mean that it consumes *free* energy, since energy as such cannot be consumed. Every activity of an organism involves consumption of free energy. The locomotion of the body as a whole uses up some of it. So does every movement of its limbs and food propulsion through the digestive canal. So does the beating of the heart, or the electric current that carries messages along the nerves and stores information in the brain. So does the maintenance of the body temperature of warm-blooded animals, since the difference between that temperature and the lower temperature of the surrounding medium represents free thermal energy.

The synthesis of the constituents of a living organism—proteins, fats, etc.—also consumes free energy. These complicated molecules may or may not contain more energy than the simpler organic molecules from which they are synthesized; but they always represent a higher degree of orderly arrangement of the atoms, and, therefore, contain less entropy. It is often said that complicated, orderly molecular systems contain a large amount of *information*. Information means, in this context, the same as negative entropy (negentropy)—a higher degree of orderliness.

A living organism is like a running clock. If it is not wound up, it will sooner or later run out of free energy and stop. If the clock of life on earth would be left to run down without rewinding, it would take less than one hundred years for all life on the planet to approach its end. First, green plants would die from starvation. Man and other animals who feed on plants would follow. And finally, bacteria and fungi feeding on dead animal and plant tissues would exhaust their food and die too.

PHOTOSYNTHESIS AND RESPIRATION

What is it that winds up the machinery of life on earth? The answer was given in Chapter 1: it is *photosynthesis*. As mentioned in Chapter 2, some terrestrial sources of free energy do exist—the heat of volcanoes or hot springs, the radioactivity of certain elements in the earth's crust. These sources, however, are rare and diffuse. The universal distribution of life on earth is bound to the infinitely more abundant extraterrestrial supply of free energy—the light energy streaming to the earth from the sun. The photosynthesis of plants is the process by which light winds up the clock of life.

This description is more sophisticated than that of Robert Mayer, quoted in the first chapter. Mayer saw in photosynthesis by plants the confirmation, on a giant scale, of the *first* law of thermodynamics: the energy of life is derived from the energy of light. The second law of thermodynamics was not known at that time, and Mayer did not ask himself "why could not life exist on earth by continuously consuming the heat energy of the oceans and the atmosphere, and converting it into chemical energy?" The second law says that this is impossible. Living organisms cannot violate the entropy principle and create order out of disorder without a compensating process, so that the net result is an increase in entropy.

The free energy needed for life is provided in all organisms—except certain bacteria—by a reversal of photosynthesis. One often speaks of "energy content of organic matter" (meaning by this its combustion energy). This is a sloppy expression. The combustion energy is the energy available not in organic matter as such, but in the system (organic matter + oxygen). In order to supply energy for life processes, organic matter and oxygen (both originating in photosynthesis) must be given a chance to react back, reversing photosynthesis. This back reaction is *respiration*—a slow, regulated combustion. Life without oxygen, that is, anaerobic life, is possible only for certain microorganisms able to extract energy from chemical processes not requiring oxygen. Thus, yeast cells can use the energy of alcoholic fermentation. Certain cells in higher organisms, such as the muscle fibers, can obtain, in emergency, some energy in a similar fashion, by fermentation leading to lactic acid.

As respiration substrates, organisms can use either materials derived from food (animals), or materials they had themselves synthesized from carbon dioxide and water (plants). The latter process is called *endogenous respiration*. However, organic materials supplied from the outside can also be used by many plants. (Carnivorous plants can even digest insects!) This is called *exogenous respiration*. Only exogenous respiration is possible in animals.

In its overall chemical result, respiration is the reversal of photosynthesis. The two processes are different in that one represents an uphill transport of material, requiring massive supply of free energy from the outside, while the other runs downhill and needs no such supply.

The energy of light acts in photosynthesis similarly to the electrical energy operating a pump that lifts water into a higher reservoir (Fig. 3.2*a*). In Fig. 3.2*b*, the accumulated chemical energy is prevented from dissipation by the metastability of the system "organic matter plus oxygen." This system, while intrinsically unstable, does not react spontaneously—even a piece of dry wood needs a match to start it burning! In organisms, the back reaction goes through special channels (Figs. 3.2*b* and 3.3), the locks on which are opened by lockkeepers—biological catalysts (enzymes). These catalytic proteins permit the release of the accumulated energy in gradual steps, rather than in a single big rush.

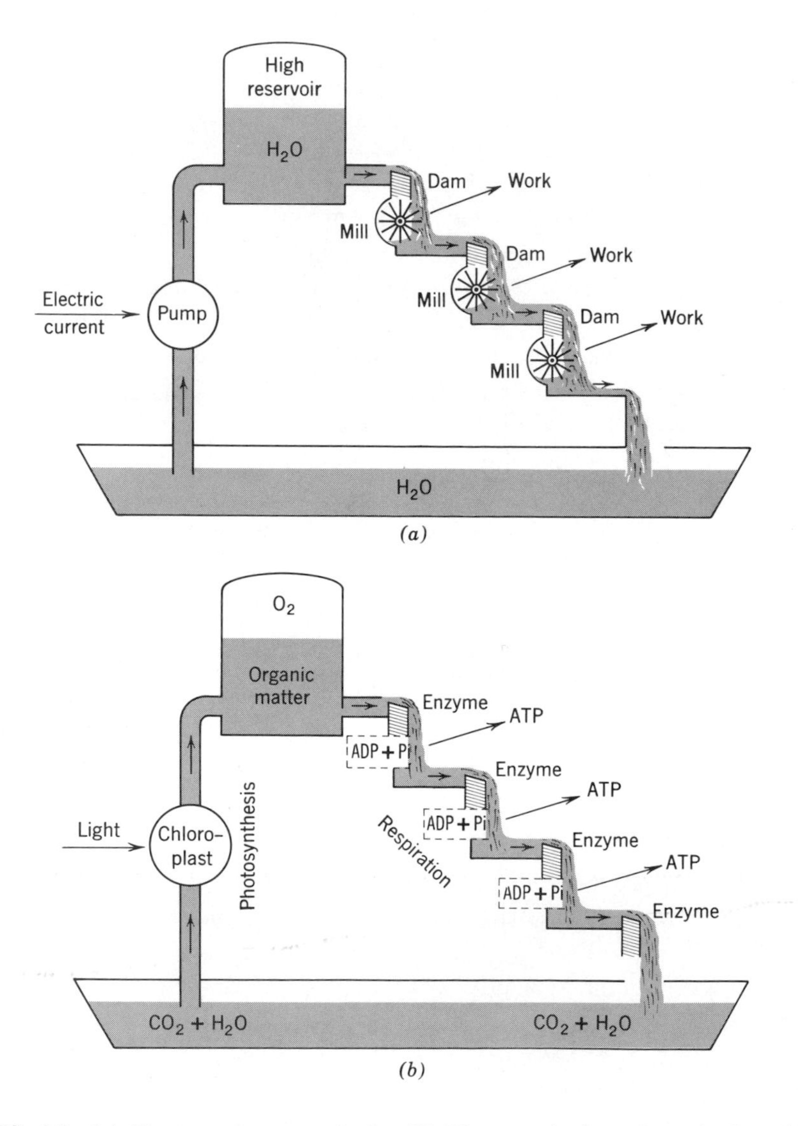

FIG. 3.2 (*a*) Pump and water wheels. (*b*) Photosynthesis and respiration. (ATP, adenosine triphosphate; ADP, adenosine diphosphate; Pi, inorganic phosphate.)

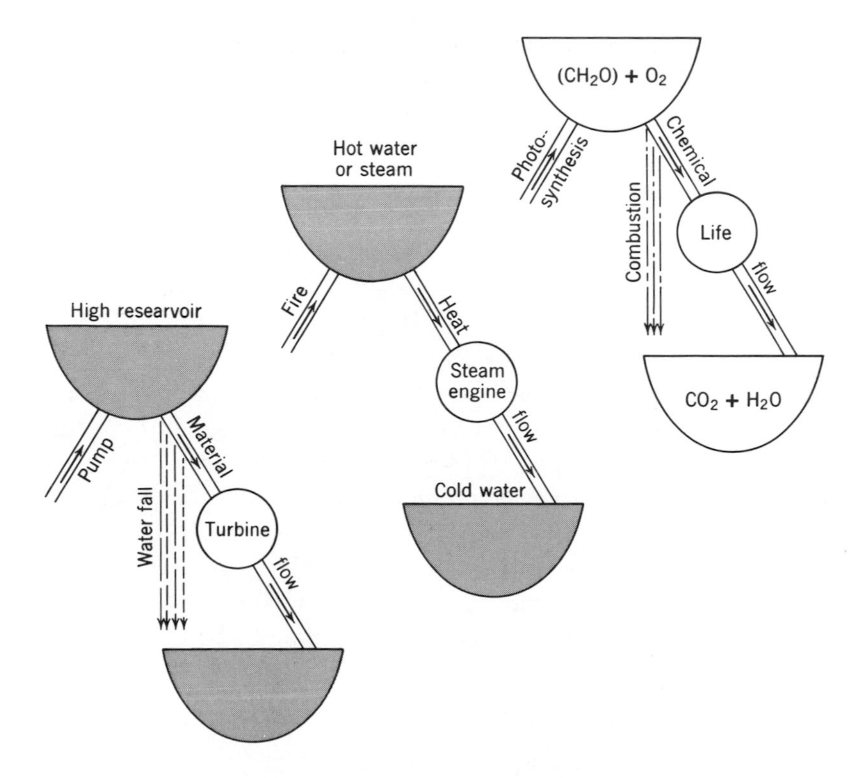

FIG. 3.3 Turbine, steam engine, life.

Such a rush would be equivalent to direct combustion (see Fig. 3.3), and even if it would not set the organism on fire, it is likely to damage it *locally*, and much of the released energy would go to waste. For its different chemical and physical functions, the organism requires small, measured amounts of energy. The respiration mechanism is fulfilling the task of releasing the free energy of the system (organic substrate + oxygen), in proper small units.

The most important energy currency in biology are the energy amounts carried by the so-called "high energy phosphate bonds" (see Chapter 18). Its most common form is that present in adenosine triphosphate, abbreviated ATP. This molecule releases about 8–10 Kcal/mole upon hydrolysis into adenosine diphosphate, ADP and inorganic phosphate, Pi.

$$ATP + H_2O \rightarrow ADP + Pi + 8\text{–}10 \text{ Kcal/mole} \qquad (3.7)$$

The "phosphate bond energy" is less than 10 percent of the energy accumulated in the elementary process of photosynthesis. The energy bank of life takes in energy in 100 Kcal bills and spends it in 10 Kcal coins (Fig. 3.2b).

It is worth pointing out here that a natural process exists which is, physically speaking, closer to the true reversal of photosynthesis than respiration. This is bioluminescence. Many animals (fishes, crustaceans, worms), and some plants (dinoflagellates, fungi, bacteria) emit light, the energy of which is supplied by a special form of oxidative metabolism. In these organisms, a certain part of the "energy capital," originally accumulated by photosynthesis, is prevented from being broken up into the "small change" of phosphate quanta, and liberated in bulk, in the form of visible light quanta of 40–60 Kcal/einstein each. (An einstein is a mole of quanta.) Since these organisms do not absorb visible light, it is not the case of absorbed quanta being reemitted as such, as in fluorescence (see Chapters 10 and 15). Rather, the light quantum becomes available by following a special pathway of oxidation—rather like a ski jump built on the down slope of a hill (Fig. 3.4).

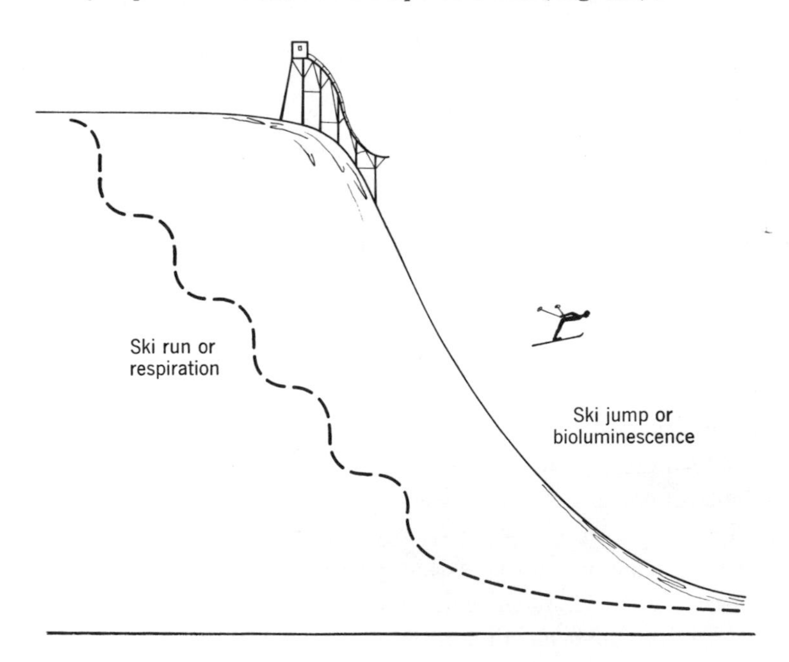

FIG. 3.4 Ski run and ski jump; respiration and bioluminescence.

ENERGETICS OF BACTERIAL PHOTOSYNTHESIS

Bacterial photosynthesis is represented by Van Niel's equation (see Eq. 2.2):

$$CO_2 + 2H_2A + \text{light} \xrightarrow{\text{bacteria}} CH_2O + 2A \tag{3.7}$$

In this equation, the amount of energy stored depends on the nature of A. We saw it to be 112 Kcal when A is oxygen, that is, in green plant photosynthesis. In the case of A being sulfur, the energy storage is about 7 Kcal, if sulfur is oxidized to sulfate. With some reductants, no energy storage occurs at all. For example, if H_2S is oxidized only to elemental sulfur, 5 Kcal of chemical energy are *lost*. In the oxidation of H_2 to H_2O (in "hydrogen" bacteria), as much as 25 Kcal are dissipated. (However, ΔF, the free energy change, is even in this case slightly positive.)

Chapter 4

Solar Energy and Its Utilization

SOLAR ENERGY SUPPLY ON EARTH

Let us now look closer at the role of solar energy in the energy household of life. At the outer boundaries of the earth's atmosphere, about 2.0 cal of energy (the amount of energy sufficient to heat 2 g. of water by 1°C) strikes, every minute, one square centimeter of the earth's cross-section. This is the so-called "solar constant"; its value means that about 1.25×10^{24} cal (1.25×10^{21} Kcal) of solar energy are received annually by the earth as a whole. Only about 40 percent of this energy, or 5×10^{20} Kcal, reaches the surface of the earth. The rest is either absorbed by the atmosphere (by oxygen, ozone, carbon dioxide, water vapor, and dust) or scattered into space. Together with radiation reflected by the surface of the earth—particularly where it is not covered by vegetation—this scattered light makes the earth shine like a star for an extraterrestrial observer, in the same way a dust particle shines in a beam of sunlight, or the moon and the cold planets shine at night for a terrestrial observer.

Despite these losses, if an economic method could be found to catch, store, and utilize the sunlight falling on our roof, it could easily cover all our domestic energy needs, while sunlight falling on large open areas could easily run all the wheels of human industry. Many attempts have been made to concentrate the energy of sunlight, for example, by means of giant concave mirrors, in order to utilize it for industrial purposes; but only minor economic successes have been achieved so far.

Concentration is needed if light is to be first converted to heat, because, according to the second law of thermodynamics, free energy can be extracted from heat only to the extent to which a temperature *difference* can be maintained; the greater this difference, the more work can be derived from it.

No such restriction applies to direct conversion of light energy into forms of energy other than heat, for example, into *electricity*. But a cheap and efficient method of such conversion has not yet been invented. Expensive devices of this kind, with a relatively high conversion yield, (of the order of 10%) are the so-called "solar batteries." They have been used as energy sources in satellites and spaceships. But one cannot economically cover large roof surfaces or land areas with highly purified silicon or other similarly expensive materials! And, so far at least, only such materials have been found suitable for the construction of solar batteries.

It is also theoretically possible to convert light energy directly into *chemical* energy, but as yet no effective and cheap way of doing this has been found. The problem is not only to find a cheap photochemical system that would store a substantial proportion of light absorbed in it; in addition, storage would have to be in a convenient form, permitting easy removal of the stored energy as needed. The answer could be a light-produced explosive mixture, a light-produced fuel, or a light-charged storage battery. None of these devices has been yet developed successfully.

To satisfy its energy needs—food, fuel, and industrial power—mankind now depends almost entirely on plants. These organisms have solved the problem of converting light energy into chemical energy with a rather low average yield, but on a vast scale. Reserves of this energy, stored in past geological areas, are available to man as fossil fuels (coal, oil, peat). Amounts currently accumulated by growing plants provide all human food (either directly, as vegetables, or indirectly, as meat or milk or animals fed on plants); a small fraction of fuel is provided in the same way, as wood or dung.

The utilization of energy stored by plants occurs by reversing photosynthesis—rapidly, in furnaces and explosion motors, or slowly, in respiring cells of plants and animals. In technologically advanced civilization, industry consumes hundreds of times more energy than human beings use up as domestic fuel. The maintenance and growth of this civilization now depends largely on the earth's accumulated capital of chemical energy (that is, on its fossil fuel reserves) and only to a very small degree on current solar energy supply, stored by plants. Mankind uses annually about 10^{13} kW hours of energy from the earth's current income, and ten times more, about 10^{14} kW hours, from the earth's accumulated capital.

The plants store less than one percent of the total solar energy reaching the surface of the earth. Man could easily live from this energy *income* of the earth, if he were able to improve significantly on the plants' ways to store it. One possible approach is breeding more efficient plants; another is growing existing plants in a way that would increase their natural rate of energy storage; the third (mentioned before) is developing nonliving systems for solar energy storage, as effective—and cheaper!—than the present-day solar batteries.

Not so long ago the need for mankind to learn how to live within its income, as far as energy consumption is concerned, seemed truly crucial. This was so because fossil fuels were then the only known significant terrestrial energy reserves, and these were being used up at an alarming and ever-increasing rate. It seemed that a great crisis of industrial civilization, if not a threat to man's very survival, was only a few centuries away. The crisis has been postponed for several centuries by the discovery of nuclear *fission*. Its energy, too, is derived from fossil fuels of a kind—uranium and thorium ores. The supplies of these fuels is also limited; but they promise to last ten or twenty times longer than the supplies of coal and oil. Finally, if *thermonuclear* energy production, through fusion of heavy hydrogen to helium, proves technically and economically feasible—as we have reason to hope it will—man's fuel reserves will become practically inexhaustible, because heavy water, D_2O, forms about 0.015% of the ocean. Nevertheless, it is reassuring to know that fundamentally, mankind could live on earth without recourse to fossil fuels of any kind, nuclear or chemical, by utilizing the flow of solar radiation. Therefore, the study of methods for effective and economic conversion of solar energy into chemical or electrical energy should not be abandoned, despite the more glamorous prospects of nuclear fission and fusion.

It has been suggested, particularly by *Farrington Daniels*,[1] America's most persistent proponent of solar energy utilization, that solar and nuclear energy could complement each other. Nuclear energy stations must be large to be economical—the larger the better!—while solar energy may be most economical where cheap, small local units, such as solar heaters, solar cookers, or solar pumps, are needed. (The use of solar power in spaceships is in a different category because cost is no consideration there.)

In any case, it seems an intellectually intolerable state of affairs! Plants, from great trees to microscopic algae, are busily engaged all around us in converting light energy into chemical energy, while man, with all his knowledge of chemistry and physics, cannot imitate them. Not only are we unable to imitate photosynthesis *as a whole* (that means, both synthesis of organic material *and* storage of energy), but we have not yet found any alternative method to convert economically the energy of solar radiation into useful chemical or electrical energy.

Nature offers man, in addition to nuclear energy, some other energy sources, not derived from photosynthesis, present or past. The most important one, in the present state of civilization, is the energy of *falling water;* but it now accounts for only about 10^{12} kW hours annually, or about one percent of mankind's total energy consumption, and the possibility of its expansion is limited. In Europe and North America, more than one half of available water power already has been harnessed. Much more untapped water power remains in Asia. Russians are now busy harnessing the big rivers flowing northward into the Arctic Ocean, the Chinese struggle with the giant but unruly rivers running east into the Yellow Sea, and India has begun the construction of hydroelectric power plants on rivers flowing into the Indian Ocean. Large, unused sources of water power remain only in South America and, above all, in Africa, where the Aswan Dam on the Nile and the Volta Dam in Ghana promise to open new regions for industrialization. However, the total water power available for development on earth is limited, and while it may be important for industrial development of certain areas, it would not begin to solve the industrial energy problems of mankind as a whole.

It is worth noting that the energy of falling water also is derived

[1] See Farrington Daniels, *Direct Use of Solar Energy,* Yale University Press, 1964.

from solar radiation. Sun heat evaporates water from the seas and thus causes rain. Rivers, fed on rain water, flow back into the sea and provide water power. The same is true of a large part of wind energy—the sun heats the atmosphere nonuniformly and thus causes winds. Use of wind energy for power is very old, but remains restricted—as all familiar with windmills in Europe and wind generators on American farms are aware.

Tidal energy, nuclear energy, volcanic heat, and wind energy associated with the rotation of the earth (rather than with its heating by the sun) are the only sources of energy on earth *not* derived from sunlight. Enormous amounts of *wind energy* go to waste, because of the difficulty of harnessing it for practical purposes. *Tidal energy* is being used, at present, only in one installation located near St. Malo in France. The use of *nuclear energy* is growing rapidly; it already accounts for about 40% of the current electric energy supply in England. About one half of newly installed electric generating power in America was based, in 1966, on nuclear fuel. *Volcanic* heat is as yet put to industrial use only in a few localities on earth, for example, in Italy.

GREEN PLANTS AS CONVERTERS OF SOLAR ENERGY

Vegetation covers all continents except the ice-coated Antartic and the deserts of Africa, Asia, and Australia. In the tropics and subtropics, vegetation utilizes sunlight all year round. In the higher latitudes, the activity of plants is reduced in winter. With the exception of the icy Arctic waters and other unproductive areas, the upper layer of water in the oceans also abounds in plants, predominantly free-swimming microscopic algae. Geochemists call the layer of organisms covering the earth the "biosphere," as if it were a thin spherical shell located between the lithosphere (earth's crust) and the hydrosphere (the world ocean) on the one side, and the gaseous atmosphere on the other. The biosphere is, however, not continuous. On land, it is interrupted by vegetation-free desert areas; in the ocean, it is not a separate layer, but is stirred up with the upper part of the hydrosphere, extending to a depth of several hundred feet under the surface—as far as light can penetrate into water. Dead organisms, plants or animals, sink to the bottom of the sea, and bacteria and some other deep-water organisms live in lightless deeps by feeding on their bodies. Despite these limitations, the term biosphere is graphic and useful.

The total amount of solar energy stored annually in the biosphere is given by the product: $A \times B \times C$ Kcal, where A is the solar energy flux hitting the earth's surface, B the percentage of this flux absorbed by plants, and C the percentage of absorbed light converted to chemical energy, with $A = 5 \times 10^{20}$ Kcal/year, $B =$ about 0.3,[2] and C about 0.01, the total is 1.5×10^{18} Kcal annually. However, the use of the same factors for plants on land and in the ocean is very uncertain.[3] If we make the calculation for fertile land areas only (that is, for about 20% of the earth's surfacc), wc obtain a figure of 3×10^{17} Kcal stored annually, corresponding to 3×10^{10} tons of synthesized organic carbon. This value is somewhat larger than that given in Table 2.1, based on land crop estimates (about 2×10^{10} tons, after correction for respiration) but it agrees with it in the order of magnitude, which is as good as could be expected.

The *maximum* efficiency of energy conversion of which plants are capable under the most favorable conditions is much higher than the above-mentioned *average* efficiency. The *minimum number of light quanta* (atoms of light, also called photons) needed to reduce one molecule of carbon dioxide and liberate one molecule of oxygen, has been much studied because of the importance of this constant for the understanding of mechanism of photosynthesis. It will be discussed in Chapter 11. The most likely figure at present is eight quanta per molecule of liberated oxygen. Eight quanta of red light carry about 350 Kcal/einstein (one einstein of quanta being numerically equal to one mole—that is, 6.0×10^{23} atoms). The maximum conversion efficiency is thus $^{120}/_{350} = 0.34$. A conversion yield of one percent, suggested above as average under natural conditions, corresponds to as many as 280 quanta per

[2] Over 50% of the solar light energy striking the earth is in the infrared spectral region, not absorbed by photosynthetic pigments. Of the remaining 50%, only about 60% is likely to be absorbed by the plants, rather than reflected from the surface of the ocean, sands, rocks, or the plants themselves.

[3] See Chapter 2 for considerably lower estimates of the average energy utilization in the ocean.

oxygen molecule of which all but eight are wasted as far as photosynthesis is concerned. (They do, however, contribute to another plant function—transpiration.)

Of the total 2–3 $\times 10^{17}$ Kcal stored by land plants each year, man utilizes as food about one percent, or 2–3 $\times 10^{15}$ Kcal annually (about 2×10^3 per day for each of 3×10^9 people on earth).

Storage of light energy as chemical energy through photosynthesis can be envisaged as a stage interpolated in the conversion into heat of the solar energy striking the earth. A steady state is established, in which a certain amount of light energy, received by the earth, is channeled off to run the "mills of life," before it is degraded into heat. Sometimes, as in the formation of coal and oil, this temporary storage of sunlight is extended over millions of years, by spatial separation of synthesized organic matter from oxygen under the cover of minerals, or under a salt dome. Sooner or later, most of the energy stored in fossil organic matter (plus atmospheric oxygen) will be dissipated by men who dig coal from under the protective layer of rock and burn it in furnaces, or pump oil from under the salt domes and use it in internal combustion engines.

From the point of view of energy transformation, the position of life in nature is similar to that of a mill built on a spillway, parallel to a waterfall (see Figs. 3.2 and 3.3). It interrupts the spontaneous process of energy degradation, represented by the conversion of light to heat or by water freely streaming downward, and diverts a part of the free energy of this degradation to useful purposes. Plants are chemical factories in which organic products of high chemical energy content are synthesized. These products are, in principle, unstable in the presence of oxygen. They perform their transformations, which are the essence of life, like acrobats on a tightrope, high above ground, surrounded by an ocean of air that continuously threatens them with annihilation. Living organisms manage to regulate the rate of this annihilation process, thus providing measured amounts of energy needed for their activities. When they die, the annihilation is completed by scavenger organisms utilizing the residual chemical energy of dead bodies for their own living processes.

Light energy is stored in photosynthesis as free chemical energy. The latter is then used by organisms for their various life activities. The *energy conversion* cycle, represented by photosynthesis and respiration, is thus coupled with a giant *chemical cycle* in which carbon dioxide and water are converted into organic matter and oxygen, and the latter back into carbon dioxide and water (see Figs. 2.1–2.3, 3.2 and 3.3). Water is the "working fluid" that mediates the transformation of heat energy into mechanical energy in a steam engine; CO_2 and H_2O are "working fluids" that mediate the transformation of light energy into life energy in plants. Water and carbon dioxide are exceptionally stable chemical compounds. They represent the most stable form in which the three elements, carbon, hydrogen, and oxygen can exist together, the lowest possible energy level of the working fluid, like cold water in the condenser of a steam engine or water in a reservoir downstream from a hydro-electric station. All useful free energy has been extracted from the system in this exhausted state. The correlated high-energy state is that of organic matter and free oxygen—similar to that of water behind the upstream barrage, or of hot steam in the boiler (Fig. 3.3).

The carbohydrates synthesized by photosynthesis are only in part utilized directly for the reverse, energy-supply process of respiration. Another part serves as raw material for a variety of chemical transformations, of which polymerization and amination (that is, introduction of nitrogen into the organic molecule) are the most fundamental ones. Some products of these transformations (cellulose, bone, cartilage, etc.,) become permanent structural parts of the organisms. Other products (fat, starch, etc.) serve as storage material, and still others (proteins, nucleic acids, vitamins) play an important role as catalysts in metabolism. Ultimately, all of them are degraded back into carbon dioxide and water, either in the normal course of life, or in decay after death.

Chapter 5

Energetics of Photosynthesis: A Closer Look

MOLECULAR ORIGIN OF PHOTOSYNTHETIC ENERGY

Why does the system "organic matter plus molecular oxygen" produced by photosynthesis contain much more energy than the raw material of photosynthesis, the system "carbon dioxide plus water"?

The two systems consist of the same atoms, C, H, and O. Each of these atoms is an assembly of positive nuclei and negative electrons swarming around them. When two different or identical atoms approach each other, electrical interactions result—attractions between positive and negative particles and repulsions between particles of the same sign. Overall attraction means that the electrons of the two atoms can arrange themselves more comfortably around the two nuclei held together than around the two separate nuclei. Overall repulsion means that the two electronic systems are most stable when the two nuclei are far apart.

The energy liberated or consumed in a chemical reaction is determined by the change of stability of the systems participating in it. In photosynthesis, we begin with a very stable arrangement of the atoms, C, H, and O, in the molecules CO_2 and H_2O, and end up with a much less stable arrangement of the same nuclei and electrons in the system $(CH_2O) + O_2$. The main reason for this decrease in stability is the fact that oxygen atoms are little attracted to each other. The molecule O_2 is a relatively loose one; its two oxygen atoms much prefer to be bound to carbon or hydrogen atoms.

The energy *liberated* when two atoms (or atom groups) A and B become attached to each other is called *bond energy*. It is *negative,* because, when the bond is formed, the energy of the system $A + B$ decreases.

In the first approximation, one can consider the total energy of a molecule as a sum of the energies of a number of atom-to-atom bonds, neglecting the much weaker "cross interactions" between atoms *not* forming bonds. For example, the energy of the molecule H_2O can be treated as the sum of the energies of two O—H bonds. The energy of each of them is then —110 Kcal/mole—one half of the total binding energy of one atom of O and two atoms of H (—220 Kcal/mole) (Fig. 5.1). As mentioned above, the "cross interaction" between the two H atoms is neglected in this approximation. The total binding energy of CO_2, —380 Kcal/mole, can be similarly considered as the sum of the energies of four C—O bonds, since each oxygen is bound to carbon by a double

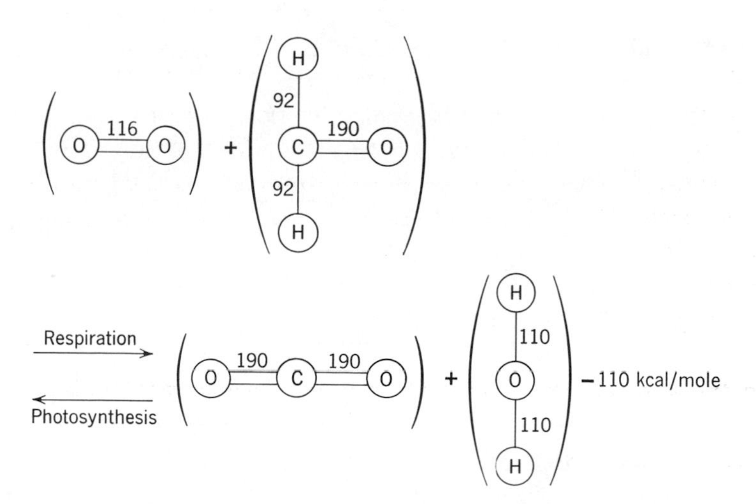

FIG. 5.1 Bond transformation in respiration and photosynthesis. The numbers are bond energies in Kcal/mole. They add up to 490 on the left and to 600 on the right side. The difference (about 110 Kcal/mole) is the energy stored in photosynthesis and liberated in respiration (see Eq. 2.1).

bond, C═O. The interaction energy of the two oxygen atom is neglected, and each single C—O bond is assigned the energy of —95 Kcal/mole.

The energy of each of the two O—O bonds in the oxygen molecule O═O is only —58 Kcal/mole (the total binding energy being —116 Kcal/mole). Therefore, every time an oxygen molecule is consumed by oxidation of an organic compound (in respiration or combustion), a considerable amount of energy is liberated by the more stable attachment of oxygen atoms to their new partners, carbon and hydrogen. This is the main reason for the large energy storage in photosynthesis, where O—O bonds (bond energy, —58 Kcal/mole) are *formed,* and O—H bonds (bond energy, —110 Kcal/mole) are *destroyed.* Other bond changes also occur in photosynthesis, such as replacement of O—H bonds by C—H bonds; but these contribute considerably less to the net energy change than the replacement of O—O bonds by C—O and O—H bonds (Fig. 5.1).

All organic matter on earth is surrounded by a swarm of free oxygen molecules, like maidens in a castle wooed by a host of suitors. Only the castle walls, the activation barriers that have to be overcome for oxidation to get underway, assure the precarious existence of living matter in contact with air. The respiration enzymes stealthily open little doors in the activation walls, and lead the organic molecules, one by one, into the embrace of oxygen.

One can formulate an approximate relation between the oxidation or combustion energy of an organic molecule and the number of oxygen molecules used up in this process. For this purpose, we define an *average reduction level, R,* of carbon in a molecule with the general composition $C_cH_hO_o$ (such as a carbohydrate, $C_nH_{2n}O_n$):

$$R = \frac{c + 0.25h - 0.5o}{c} \tag{5.1}$$

R is the number of oxygen molecules needed to burn a compound to CO_2 and H_2O, divided by the number of C-atoms present in the molecule. (Each C-atom requires one molecule of oxygen to be converted to CO_2; each H-atom requires one quarter of a molecule of oxygen to be converted to H_2O, and each atom O, already present in the molecule, diminishes by one-half molecule the number of outside O_2 molecules needed for combustion.) The approximate rule is that the *energy liberated in combustion of a molecule* $C_cH_hO_o$ *(its "heat of combustion") is about 110 Kcal/mole per carbon atom per unit* R.

For a carbohydrate, (CH_2O), R is equal to 1.0; while for the fully reduced one-carbon molecule, CH_4 (methane), $R = 2$. Photosynthesis thus lifts the raw material, CO_2 ($R = O$), roughly halfway up the range of reduction levels, and thus stores about one half of the maximum possible combustion energy per atom of carbon. (Actually, it is somewhat *more* than one half, because the plot of the heats of combustion as function of the reduction level is slightly concave. The energy storage slows down as the reduction progresses; the combustion energy of methane is 211 Kcal/mole, rather than $2 \times 110 = 220$ Kcal/mole.)

After photosynthesis has lifted carbon dioxide about halfway up the reduction scale, two kinds of follow-up processes can occur. These are *oxidations,* with the liberation of the stored energy, as in respiration; and organic *syntheses,* involving transformations of carbohydrates into other organic compounds, such as proteins, fats, etc. To the extent to which some of these compounds, particularly fats and oils, have a higher reduction level than the carbohydrates, their synthesis requires additional energy. (For example, R is about 1.4 for a simple fat, such as triglycerate of oleic acid, $C_{57}H_{104}O_6$.) Nevertheless, these syntheses do occur in the dark, without the supply of light energy. How is this possible? Nature does this by coupling an *oxidation,* which liberates energy, with an energy-accumulating *reduction,* making the net free energy change negative. Reactions of this type are called *dismutations.* The metabolic processes known as fermentations belong to this class. In alcoholic fermentation of glucose, $R = 1.0$, some of it is *reduced* to ethyl alcohol, C_2H_5OH ($R = 1.5$), while another part is *oxidized* to carbon dioxide ($R = O$). The net effect is the release of only 21 Kcal/mole, as compared with 670 Kcal/mole released when the same molecule is oxidized by O_2 to CO_2 and H_2O in respiration.

Processes in which the reduction level remains unchanged, such as conversion of one sugar into another, polymerization of a sugar to starch or cellulose, hydration (that is, addition of water), and dehydration (loss of water) by an organic molecule, proceed on an almost constant level of energy.

Here, then, is one way to look at the multitude of synthetic and degrading processes in the organic world. Photosynthesis lifts the reduction level of carbon from 0 to 1, and respiration brings it back to zero.

Other metabolic transformations keep the *average* reduction level constant; if one part of a substrate is oxidized, the other part is reduced.

OXIDATION-REDUCTION POTENTIALS

It is useful to introduce here another way of describing the energetics of oxidation and reduction processes. The original meaning of the term "oxidation" was *addition of oxygen* (as in rusting) and the original meaning of "reduction" was *removal of oxygen* (as in the preparation of a metal by smelting of an oxide ore). In organic chemistry, this picture was replaced by that of the *loss or acquisition of hydrogen atoms.* Thus, photosynthesis, once described as subtraction of oxygen from carbon dioxide and hydration of the remaining carbon:

$$CO_2 \rightarrow C + O_2;\ C + H_2O \rightarrow (CH_2O)$$

can be better represented as transfer of hydrogen atoms from water to carbon dioxide:

$$2H_2O \rightarrow O_2 + 4H \text{ (removal of H-atoms from water)} \qquad (5.2a)$$

$$4H + O{=}C{=}O \rightarrow \left[\begin{array}{ccc} H & H & H \\ | & | & | \\ O & -C- & O \\ & | & \\ & H & \end{array}\right] \rightarrow [(CH_2O) + H_2O] \qquad (5.2b)$$

(addition of H-atoms to carbon dioxide, followed by dehydration)

which adds up to

$$CO_2 + 2H_2O \rightarrow (CH_2O) + H_2O + O_2 \qquad (5.2)$$

The dehydration of the immediate hydrogenated product, HO—CH_2—OH, indicated in parentheses in (5.2b) is of little significance energetically.

This new description permitted the inclusion, under the heading of oxidations and reductions, of processes in which oxygen takes no part at all. For example, the oxidation of a hydroquinone to a quinone

$$\text{Hydroquinone} \xrightarrow{-2H} \text{Quinone}$$

can be represented (in the case of benzohydroquinone) by:

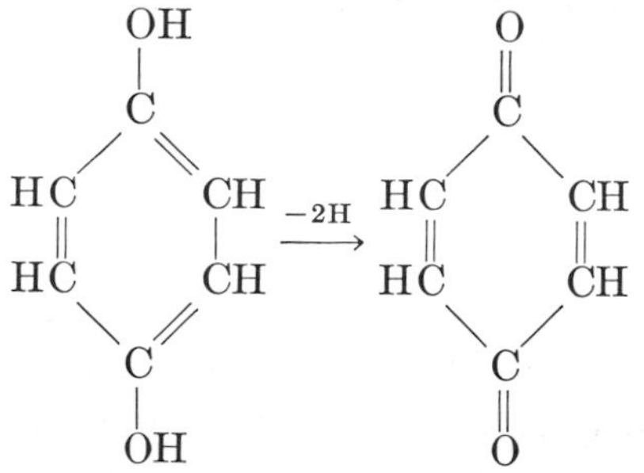

```
          OH                       O
          |                        ||
          C                        C
        /   \\                   /   \
     HC       CH    -2H      HC       CH
     ||       |     --->     ||       ||
     HC       CH             HC       CH
        \   //                  \   /
          C                        C
          |                        ||
          OH                       O
```

Generally, all processes of the type

$$AH + B \rightarrow A + BH \tag{5.3a}$$

or

$$AH_2 + B \rightarrow BH_2 + A \tag{5.3b}$$

are called reductions of B and oxidations of A. Obviously there can be no such oxidation without reduction. Hydrogen atoms are transferred from the reductant (or hydrogen donor) which is thus oxidized to the oxidant (or hydrogen acceptor) which is reduced. It is thus better to speak of "oxidation-reduction," rather than of oxidation *or* reduction. Every oxidation-reduction reaction requires two oxidation-reduction "couples," that is, two compounds that can exist in a hydrogenated (reduced) and a dehydrogenated (oxidized) state. In Eq. 5.3a the two participating couples are A/AH and B/BH. In photosynthesis, one oxidation-reduction couple is $CO_2/[(CH_2O) + H_2O]$, the other, O_2/H_2O. Photosynthesis is thus a reduction of CO_2 to CH_2O, and an oxidation of H_2O to O_2—however strange it may seem at first to call oxidation a process that leads to *liberation* of oxygen!

All oxidation-reduction ("redox") couples can be arranged on a linear scale, placing on top the strongest oxidants and weakest reductants, and at the bottom the strongest reductants and weakest oxidants. The couple $O_2/2H_2O$, or—if we use Eq. 5.3b—$\frac{1}{2}O_2/H_2O$, will be near the top, since oxygen is a strong oxidant. The couple $CO_2/[(CH_2O + H_2O)]$ will be low on this scale, since sugar is a fairly strong reductant.

A given redox couple can be expected to oxidize any couple below it on the scale and to be oxidized by every couple above it. This, however, is only true in the first approximation, because the direction of a chemical reaction depends on the sign of the change in *free* energy, and not that of the change in *enthalpy*. The change in free energy depends not only on the *nature* of the participating and resulting compounds, but also on their relative *concentration*. A weak oxidant, present in a very high concentration (relative to that of its reduced form), will oxidize the reduced form of a stronger oxidant, particularly if the latter is present in a very low concentration compared to *its* reduced form. To characterize unambiguously the relative oxidative and reductive powers of different redox couples, they are usually compared with each other under the so-called "normal conditions," that is, when the two forms, the reduced and the oxidized, are present in equal concentrations.

Having made the step from defining oxidation and reduction as acquisition or loss of *oxygen*, to defining it as loss or acquisition of *hydrogen* atoms, we now make a second step and generalize the concept of oxidation and reduction still further, by defining them as loss or acquisition of *electrons*. A hydrogen atom can be considered as consisting of an H^+-ion and an electron, $H = H^+ + e$. It is the addition or subtraction of the electron that matters in oxidation-reduction. The addition or subtraction of an H^+-ion does not contribute to it; it defines a different kind of reaction, the *acid-base* transformation, such as $H_2SO_4 \rightarrow H^+ + HSO_4^-$, or $HCl \rightarrow H^+ + Cl^-$. The general equation of a redox reaction is, according to this new definition:

$$A^- + B \rightarrow B^- + A \text{ (electron transfer from A to B)} \quad (5.4)$$

For ease of presentation, the reductant in Eq. 5.4 was assumed to be a negative ion, A^-; but it may as well be a neutral atom, or even a positive ion, able to lose a further electron, as in $Fe^{2+} \rightarrow Fe^{3+} + e$.

Equation 5.4 represents the actual mechanism of some oxidation-reductions. For example, when ferrous salts, containing the ions Fe^{2+}, are oxidized by ceric salts, containing the ions Ce^{4+}, the actual reaction is

$$2Fe^{2+} + Ce^{4+} \rightarrow 2Fe^{3+} + Ce^{2+}$$

(transfer of two electrons from iron to cerium)

In other cases, the actual result may be transfer of hydrogen atoms, rather than of electrons. However, since in all aqueous systems, hydrogen ions, H^+, are present, we can interpret hydrogen atom transfers as electron transfers combined with acquisition (or loss) of H^+-ions. The general equation (5.5), for example, can be divided into steps (5.5a, b, and c).

$$AH + B \rightarrow A + BH \text{ (transfer of an H-atom)} \quad (5.5)$$

can be divided into

$$AH + B \rightarrow AH^+ + B^- \text{ (electron transfer)} \qquad (5.5a)$$

$$AH^+ \rightarrow A + H^+ \text{ (loss of } H^+ \text{ ion; base} \rightarrow \text{acid transformation)} \qquad (5.5b)$$

$$B^- + H^+ \rightarrow BH \text{ (acquisition of } H^+ \text{-ion; acid} \rightarrow \text{base transformation)} \qquad (5.5c)$$

In biological oxidation-reduction processes, some stages may be electron transfers and others, H-atom transfers. The former is the case in one important part of the reaction sequence of respiration, in which several so-called "cytochromes" act as intermediates. These are protein molecules carrying a porphyrin group with an iron atom in either the ferrous (Fe^{2+}) or the ferric (Fe^{3+}) state. The stage in respiration involving a sequence of several cytochromes is often referred to as the "electron transfer chain." A corresponding, even if apparently shorter, chain has been recently discovered in photosynthesis (see Chapters 13–16).

What makes representation of oxidations and reductions as electron losses and acquisitions particularly useful, is the possibility it offers to replace the somewhat abstract magnitudes, the *free energies* of oxidation-reduction reactions, by the (often directly measurable) *oxidation-reduction potentials* (redox potentials).

Since we will repeatedly use the term redox potential in the following discussion, some explanation is in order.

One way in which electrons can be added or subtracted from many atoms or molecules in aqueous solution is by interaction with the electrode in a galvanic cell. As an example, we can consider the so-called Daniel cell (Fig. 5.2). Placing a copper electrode in the solution of a copper salt, and a zinc electrode in the solution of a zinc salt, separating the two solutions by a porous wall to prevent mixing, and joining the two electrodes by an outside wire, we will find copper ions, Cu^{++}, being discharged and precipitated at the copper electrode, and zinc atoms going into solution from the zinc electrode as zinc ions, Zn^{++}.

In dissolving, zinc atoms leave their electrons in the electrode, from where they are conveyed, through the external wire, to the copper electrode. The result is *reduction* of copper ions, and *oxidation* of zinc atoms.

$$Cu^{++} + Zn \rightarrow Cu + Zn^{++}$$

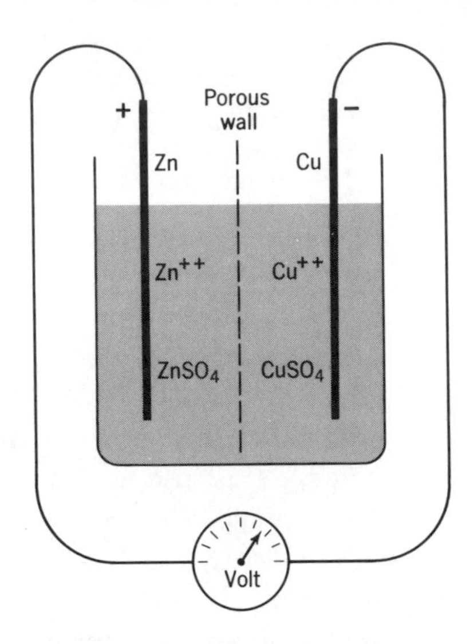

FIG. 5.2 A galvanic cell (Daniel cell). Electrons flow in the outside circuit from the zinc to the copper electrode.

The difference between the oxidative powers of the Zn^{++}/Zn and Cu^{++}/Cu couples is given directly by the galvanic potential of the cell, that is, by the counterpotential that has to be applied to *prevent* the electrons from moving from zinc to copper in the outer circuit. This potential can be measured by a voltmeter. One can plot the so-determined *oxidation-reduction potentials* (redox potentials) of different redox couples, instead of their free energies of oxidation-reduction, on a linear scale (Fig. 5.3) and use them to characterize the relative strength of oxidants and reductants. Of course, these potentials, like the corresponding free energies, will depend on the *ratios of the concentrations* of the oxidized and the reduced forms. We define as *normal* potentials, E_0, those obtained when these concentrations are equal.[1] Furthermore, if the redox reaction is associated with a loss or acquisition of H^+-ions, the potential will depend on the concentration of the latter (the pH of the solution[2]). The normal potential in *neutral* solution (pH7) is often denoted by E_0'.

Many, in fact most, organic oxidation-reduction couples are "electrode

[1] In the case of one form being a solid or liquid (rather than a solution), only the other one has to be present in unit concentration; in the case of one form being a gas, its normal condition is one atmosphere pressure.

[2] pH is defined as the negative logarithm of the H^+-ion concentration; $pH = -\log [H^+]$.

inactive," so that their oxidation-reduction potentials cannot be determined in a galvanic cell. In this case, to put them in proper position in Fig. 5.3, one must return to the determination of *free energy* change (ΔF), and calculate the potential difference (ΔE), by multiplication with an appropriate proportionality factor, F, according to the equation

$$\Delta F = nF\,\Delta E \tag{5.6}$$

where n = number of electrons moved, and F the so-called Faraday's constant (the total electric charge carried by a mole of electrons).

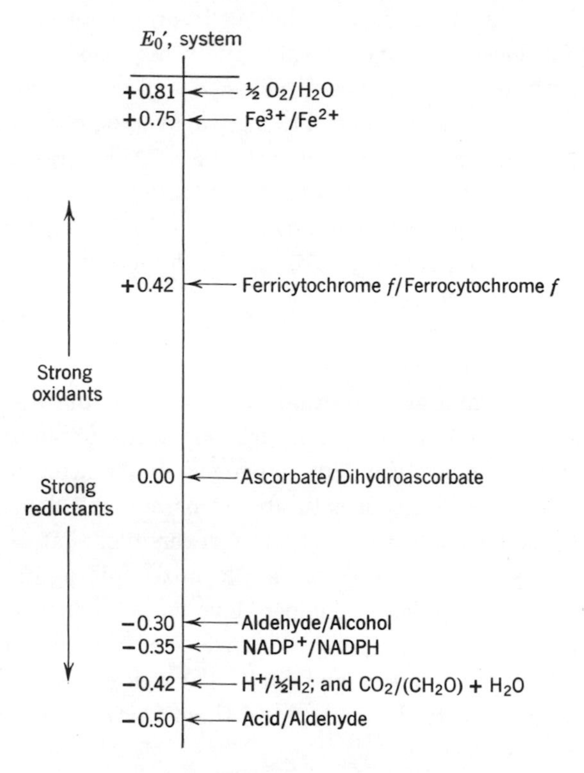

FIG. 5.3 Scale of normal redox potentials (E_0') (in volts; in neutral solution, pH7). The potentials of the couple $H^+/\frac{1}{2}H_2$ at pH0 (i.e., in one-normal acid) is arbitrarily set at zero; this couple has a potential of −0.42 volt at pH7. $NADP^+$ and NADPH mean oxidized and reduced adenine dinucleotide phosphate, respectively. The calculated redox potential of the couple carbon dioxide/carbohydrate is approximately −0.40 volt. The potential of the couple ferricytochrome *b*/ferrocytochrome *b* is −0.40 volt, and that of the couple ferricytochrome *c*/ferrocytochrome *c* is +0.26 volt.

Speaking of the normal potential of a given redox couple is embarrassing; we said above that each oxidation-reduction must involve *two* couples! In fact, the only "real" thing is the difference between the two redox potentials. It is, however, convenient to define—quite arbitrarily—the oxidation-reduction potential of a certain standard couple as zero. One can then ascribe single potentials to all other couples—the potentials one would obtain by making these couples react with the standard couple in a galvanic cell. The standard is the couple $H^+/\frac{1}{2}H_2$, where H^+ is the oxidant and H_2 the reductant. This system can be realized, in practice, by bubbling hydrogen through acid solution at a platinum electrode covered with finely dispersed platinum. At such an electrode, hydrogen ions are discharged and converted into hydrogen gas smoothly (that is, without having to overcome an energy barrier, as on many other electrode-solution interfaces). The pressure of hydrogen must be one atmosphere, and the concentration of H^+ ions in solution, 1 mole/liter (a so-called one-normal solution). In neutral solution (pH7), the normal potential, E_0', of the hydrogen electrode is —0.42 volt (Fig. 5.3).

Every couple that has a redox potential more negative than that of the hydrogen electrode is thermodynamically able to evolve gaseous hydrogen from a normal acid solution, and every couple with a more positive potential than that of the hydrogen electrode should be reduced, in one normal acid solution, by hydrogen under atmospheric pressure. Whether these reactions will actually go or not will depend on whether the couple can be made to react at a reversible, that is, "frictionless" electrode, as hydrogen ions do at a platinized platinum electrode, zinc ions at a zinc electrode, or copper ions at a copper electrode. With most organic couples, this is not the case. Nevertheless, it is convenient to calculate from thermodynamic data their redox potentials with respect to the hydrogen electrode, and place them in their proper position on the redox potential scale.

In photosynthesis, the situation is as follows: the normal potential of the system $\frac{1}{2}O_2/H_2O$ is +0.81 volt (oxygen gas is supposed to be under one atmosphere pressure, pH to be 7, and water to be liquid). The potential of the system $CO_2/[(CH_2O) + H_2O)]$, calculated from thermodynamic data, is about —0.40 volt. The difference between the two potentials, 1.21 volt, is a measure of the free energy of photosynthesis.

One may object that volts are a measure of electric *potential,* not energy. In order to convert potentials into energies, one has to multiply them by the charge moved up or down over the potential difference, as the energy of falling water is obtained by multiplication of the height of the fall by the amount of water carried over it. Thus, a 110 volt lamp, through which a current of 0.5 ampere, or 0.5 coulomb/second passes, consumes 55 watt (that means 55 joules/second, joules being a measure of energy, one joule = 10^7 erg). In dealing with electrons, the usual energy unit is *electron-volt.* One electron-volt is the energy consumed or released when a single electron is moved up or down a potential difference of one volt. Thus, the scale of redox *potentials* in *volts* can be treated as a scale of redox *free energies* in *electron-volts.*

It is important to note that oxidation-reduction processes involving the transfer of several electrons or hydrogen atoms, require or liberate proportionally more energy. For example, in photosynthesis, *four* hydrogen atoms have to be transferred from H_2O to CO_2 in order to reduce the latter to the reduction level of a carbohydrate:

$$CO_2 + 2H_2O \rightarrow (CH_4O_2) = [(CH_2O) + H_2O] + O_2$$

That the compound (CH_4O_2) is not stable, but loses one H_2O molecule and becomes (CH_2O), is unimportant from the point of view of energy, because water loss does not change the reduction level; in organic chemistry, whenever two OH groups find themselves attached to the same carbon atom, the tendency is for the structure $RC(OH)_2$ to transform itself into $R—C═O + H_2O$, with little change in energy.

To calculate the free energy of photosynthesis, one has to multiply the above-quoted potential difference of 1.21 volts by a factor of 4 (for 4 hydrogens) giving 4.84 electron-volts. Conversion table for various energy units gives, for the caloric equivalent of one electron-volt, 23.0 Kcal/mole. Multiplying 4.84 by this factor, we obtain 112 Kcal/mole as an approximate free energy of photosynthesis under normal conditions. (The true standard free energy of photosynthesis, as calculated from thermodynamic considerations, is a few Kcal higher.)

We can summarize this section by saying that plant photosynthesis is an oxidation-reduction reaction between an oxidant (CO_2) with a potential of about —0.4 volt, and a reductant (H_2O) with a potential of about +0.8 volt, in which *four* electrons (or four H-atoms) are transferred "uphill," against a potential gradient of about 1.2 volts. This picture will be particularly helpful to use when discussing the likely steps in the overall processes.

We can thus represent photosynthesis by the scheme shown in Fig. 5.4 (compare with Fig. 5.3, where O_2/H_2O couple is near the top of the figure and CO_2/CH_2O couple near the bottom). The vertical arrow in this scheme shows the "uphill" hydrogen transfer from the potential level of +0.8 volt to a level of —0.4 volt. This uphill transfer of H-atoms is the essence of the energy-storing photochemical stage

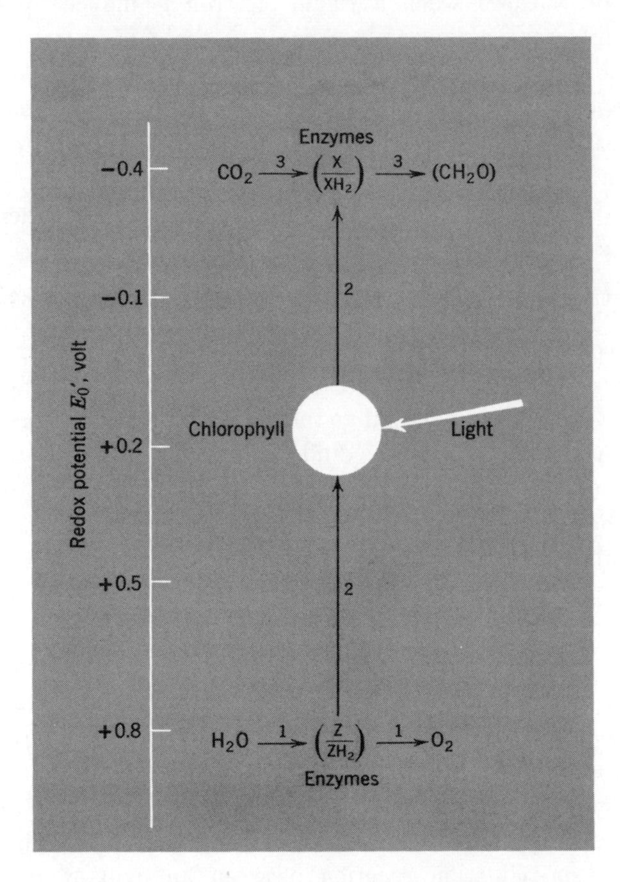

FIG. 5.4 Photosynthesis divided in three parts. (1) Enzymatic transformation of H_2O to O_2. (2) Hydrogen transfer from an intermediate (Z/ZH_2) in enzymatic sequence (1), to an intermediate (X/XH_2) in enzymatic sequence with the help of light-activated chlorophyll. (3) Enzymatic transformation of CO_2 to (CH_2O).

in photosynthesis.[3] The two series of horizontal arrows in Fig. 5.4 indicate nonphotochemical reactions occurring on an approximately horizontal (or downward-sloping) level on the energy scale. The upper arrow represents the sequence of enzymatic reactions by which the oxidant, CO_2 (or an organic molecule into which CO_2 had been incorporated by an enzyme-catalyzed reaction; see Chapter 17) is transformed into a carbohydrate. The lower arrow similarly represents the enzymatic reactions by which the reductant, water (or a compound into which water had been incorporated by a dark reaction) is converted into the final oxidation product, molecular oxygen. Figure 5.4 will serve as the basis for our subsequent discussions of the mechanisms of photosynthesis.

Bacterial photosynthesis, mentioned in Chapters 2 and 3, is an oxidation-reduction reaction in which compounds other than water serve as reductants. The amount of free energy stored in bacterial photosynthesis is much smaller (sometimes, there is even a loss of free energy) because the oxidation-reduction potentials (E_0') of bacterial reductants are much lower than that of the system $H_2O/\frac{1}{2}O_2$. For example, E_0' of the couple $H^+/\frac{1}{2}H_2$ is as low as -0.42 volt (at pH7); that of the couple H_2S/S is approximately -0.2 V, (as contrasted to $+0.8$ volt for the couple $H_2O/\frac{1}{2}O_2$).

[3] That negative levels appear in this scheme higher than the positive ones, is due to the convention, common to biologists the world over (as well as to European physical chemists), which makes strong oxidants *positive,* and strong reductants *negative;* American physicochemical convention used to assign, instead, positive potentials to strong reductants. The difference of potentials measures in one case the free energy of oxidation, in the other case, the free energy of reduction.

Chapter 6

Taking Photosynthesis Apart. I. The Light and the Dark Stage

THE PHOTOCHEMICAL AND THE ENZYMATIC STAGE

For a long time after the establishment of the overall chemical equation of photosynthesis (Eq. 1.3) and of its energy balance (Eq. 2.1), the process remained inaccessible to further analysis. In studying many other metabolic processes, for example, respiration, fermentation, and the synthesis of many important cell components, biochemists have long since learned how to extract the cells' catalytic components, the enzymes, or enzyme-bearing structural units (subcellular particles), and to carry out the pertinent reactions outside the living cell. Substrates on which these enzymes operate also could be extracted from the cells, or prepared synthetically. Ultimately, by putting together a proper assortment of substrates and enzymes, the whole metabolic process could be reconstructed in an artificial system. In recent years, this taking apart and putting together of metabolic processes has extended into the inner sanctum of life—the self-duplication of nucleic acids, which is the basis of heredity, and the synthesis of proteins, the process by which hereditary instructions, laid down in nucleic acids, are implemented as structural and functional capacities of an organism.

Photosynthesis long resisted such taking apart and reconstruction. The photosynthesizing cell appeared as a magic box, a *camera obscura,* into which nature throws carbon dioxide and water, exposes it to light, and presto, out comes oxygen, and carbohydrates accumulate inside. As soon as the cell was destroyed, or substantially damaged, this magic capacity was lost. For a long time, no partial steps of photosynthesis could be reenacted with nonliving material, not even with extracts or fragments from living cells. The mechanism of photosynthesis, therefore, long remained a matter of pure speculation.

To quote one example: In 1870, a German chemist, A. von Baeyer, suggested that the transformation of carbon dioxide in photosynthesis could proceed in two consecutive steps. First, CO_2 is *reduced* to the simplest carbohydrate, formaldehyde:

$$CO_2 \xrightarrow{+4H} CH_2(OH)_2 \; (\rightarrow CH_2O + H_2O)$$

and then formaldehyde is *polymerized* to a carbohydrate, $(CH_2O)_n$. (This hypothesis was based on the capacity of formaldehyde, observed by von Baeyer, to polymerize to "formose," a sugar-like substance.) Attempts to test this hypothesis by introducing formaldehyde into plants as a substrate for sugar production failed. (Formaldehyde is a cell poison; this is why it is used as a disinfectant.) Nevertheless, until a few years ago, textbooks of plant physiology quoted von Baeyer's scheme as the best that could be said about the chemical mechanism of photosynthesis.

The reason why attempts to break photosynthesis into partial processes have proved unsuccessful is that, in contrast to metabolic processes that proceed with a *decrease* in free chemical energy, photosynthesis involves its *storage*. This makes photosynthesis as different from ordinary metabolic reactions as pumping water up into a high reservoir is different from its running down through a series of turbines and turning wheels as it runs (see Figs. 3.2 and 3.3). This pumping-up must involve formation of highly unstable intermediates, which undergo rapid enzymatic stabilization, ending in the liberation of oxygen and formation of carbohydrates. These intermediates cannot be easily extracted to reconstruct in vitro parts of the photochemical process.

Photosynthesis apparently requires certain submicroscopic structures in the cell to prevent unstable intermediates from mutual destruction and direct them into proper enzymatic reaction channels. Something similar does exist in respiration, where a certain sequence of energy releasing reactions cannot be separated from the subcellular particles called *mitochondria*. In this case, all partial reactions occur downhill, so that intermediates could not escape by back reactions. Nevertheless, it seems important to keep them on the right track by providing a structural background that makes the reactions occur in proper order. We will deal in Chapter 8 with what is known about the structure of the subcellular particles required for photosynthesis. In the present chapter, we turn to another subject: the discovery of the division of photosynthesis into a photochemical and an enzymatic stage.

Unable to take photosynthesis apart, students of this process were in the position of mechanics wanting to interpret the operation of an automobile but not permitted to lift the hood and dismantle the engine. What could they do but resort to kinetic measurements, determining the speed with which the car runs in relation to supply and consumption of gas and air, hoping to obtain in this way some insight into the operation of the engine? Because of this situation, more numerous and precise kinetic data have been accumulated in the study of photosynthesis than in that of respiration, or of other metabolic processes more easily accessible to biochemical dismantling.

Incidentally, how does one measure the rate of photosynthesis? The earliest method was to count the oxygen bubbles rising from an illuminated submerged plant. This rough procedure was later developed into precise physical methods of measuring the amount of oxygen liberated by a suspension of unicellular algae. One such method is manometry. However, a complication appears in its use: according to Eq. 2.1, one volume O_2 is produced when one volume CO_2 is consumed. To prevent the two changes from balancing each other, the CO_2-consumption effect must be eliminated as fully as possible. This can be done by suspending the algae in a carbonate-bicarbonate mixture. Removing CO_2 from such a buffered solution merely causes the conversion of carbonate into bicarbonate, without any CO_2 being taken up from the gas above the solution. An inconvenience of this method is that it requires the use of alkaline media, which not all plants find to their liking. A more sophisticated approach permits the use of neutral or slightly acid solutions. In this method, *two* manometers are used, with a different gas volume above the suspension. The rates of exchange of CO_2 and O_2 can then be calculated separately, from the readings of two manometers, taking into account the known differences in the solubility of the two gases in water.

Other methods of measuring the rate of photosynthesis include mass spectrometry, infrared spectroscopy, calorimetry, polarimetry, and chemical analysis. The most convenient one is a variation of the polarimetric method, in which the rate of addition or subtraction of oxygen to the medium is determined by measuring the electric current flowing through it, between a platinum and a silver-silver chloride electrode. A negative potential is applied to the platinum electrode.

What organisms are most convenient for quantitative photosynthetic research? A favorite object has been the unicellular green alga *Chlorella pyrenoidosa*—somewhat like the fruit fly in the study of genetics. Another much used unicellular green alga is *Scenedesmus;* unicellular flagellates, such as *Euglena gracilis* (organisms that can live either as "animals" or as "plants") also have been used. Useful for comparative study are unicellular red algae (such as *Porphyridium cruentum*), or blue-green algae (such as *Anacystis nidulans*), and certain species of diatoms (such as *Navicula minima*).

Multicellular plants, leaves, and algal fronds, which have been the main subjects of earlier qualitative investigations, are less suitable for quantitative studies than the unicellular algae, because the latter can be suspended in an appropriate medium and stirred during the measurement to maintain uniform conditions. However, when one works with chloroplasts, the subcellular organelles in which photosynthesis occurs (see Chapters 7 and 8), higher plants can be used for preparing them. A favorite material for this purpose is spinach (*Spinacea oleracea*).

Photosynthetic *bacteria* are increasingly being used for comparative studies in photosynthesis because of the characteristic similarities and differences of their behavior compared to that of higher plants and algae. There are two main types of these bacteria—purple (such as *Rhodospirillum rubrum* and *Rhodopseudomonas spheroides*), and green (such as *Chlorobium thiosulfatophilum* and *Chloropseudomonas ethylicum*). These bacteria are abundant in stagnant, oxygen-deprived natural waters.

With recent successes in "opening the hood" of photosynthesis (see Chapters 7 and 17), interest in rate measurements has subsided. And yet, a completely satisfactory interpretation of a metabolic process would have to be quantitative, and not merely qualitative. In other words, we need to know not only the sequence of the chemical steps and the nature of enzymes catalyzing them, but also why the overall reaction runs at the actually observed rate, and responds in a certain way to changes in external conditions. Such quantitative tests can be fatal to many a qualitatively plausible hypothesis, particularly in the case of photosynthesis, where high efficiency is an essential feature of the whole process.

Kinetic evidence of two types has led to a two-step concept of photosynthesis, involving one light-requiring step and one "dark," that is, not light-requiring step: (1) measurements of the rate of photosynthesis as function of the intensity of *steady illumination* and (2) measurements of the rate of photosynthesis in *flashing light* as function of the energy of the flashes and of the duration of dark intervals between them.

LIGHT SATURATION AND ITS IMPLICATIONS

As first noted by the German botanist J. Reinke, in 1883, the proportional increase in the rate of photosynthesis, P, with increasing intensity of illumination, I, is replaced, in sufficiently strong light, by light *saturation,* as shown in Fig. 6.1. The exact meaning of "strong" depends on whether we deal with shade-loving (umbrophile) objects, such as shade leaves or deep-water algae, or with light-loving (heliophile) ones, such as sun-exposed leaves, surface algae, and, particularly, desert and alpine plants. The plants of the first type may become light-saturated at one tenth or less of full sunlight at sea level, whereas plants of the second type may not be light-saturated even in direct sunlight at noon.

After saturation, the rate remains constant over a certain range of light intensities. In still stronger light, the rate begins to decline, particularly if the illumination is prolonged. This decline is caused by irreversible injury by light; the rate is not restored after return to lower illumination. On the other hand, the rising and the horizontal part of the light curve, $[P = f(I)]$, can be reproduced again and again by going either from lower to higher, or from higher to lower intensities.

The British plant physiologist, F. F. Blackman, was the first, in 1905, to interpret the shape of the light curves of photosynthesis as evidence of a two-step mechanism, consisting of a photochemical and a dark step. For some time, the latter has been widely called the "Blackman

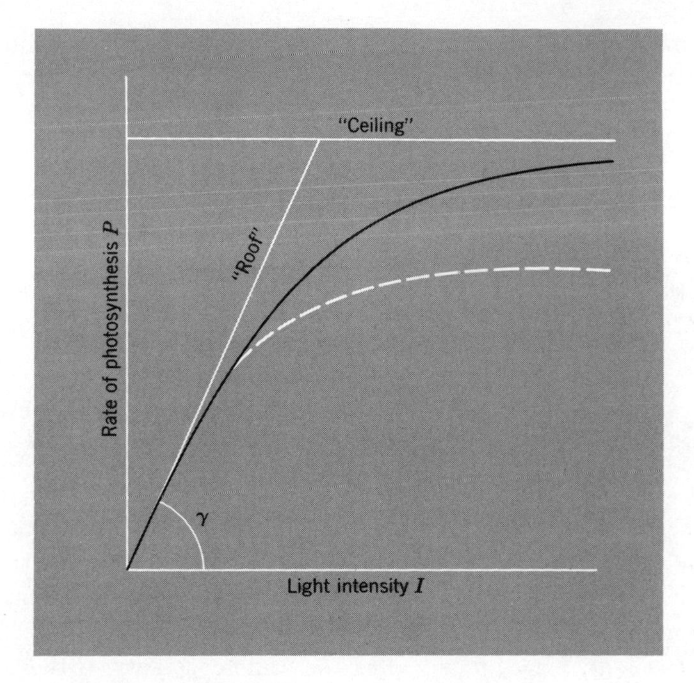

FIG 6.1 Light curve of photosynthesis [$P = f(I)$]. Dashed line shows the effect of lowering the temperature or adding a poison such as cyanide. The tangent of the angle γ measures the maximum yield of photosynthesis (provided I is the *absorbed,* not the *incident,* quantum flux).

reaction"; but this term is rarely used now, since we know that photosynthesis involves not one, but many dark enzymatic reactions.

The initial part of the light curve, in which the rate of photosynthesis increases proportionally with light intensity, corresponds, according to Blackman, to the so-called *light-limited state* of photosynthesis. In this range, as quickly as the light-produced primary photochemical products are formed, they are further transformed by dark (that is, nonphotochemical) reactions. It is the rate of supply of light that limits the overall rate under these conditions. When *light saturation* begins to manifest itself by curvature of the light curve towards the horizontal, this is evidence that the dark chemical apparatus is becoming overtaxed and incapable of taking care of all the primary light products as rapidly as they are formed.

This makes sense. However, it is worth pointing out that light saturation does not occur in "ordinary" photochemistry in vitro, even if the primary photochemical reaction *is* followed by a dark reaction. Light saturation can, in fact, occur only if the dark reaction that follows the photochemical step has a certain maximum "ceiling" rate. This is not the case in ordinary chemistry, but is typical of enzyme-catalyzed reactions in biological systems.

The general mechanism of enzyme-catalyzed reactions, first suggested in 1913 by the German biochemist, Lenor Michaelis, consists of two steps.

$$S + E \rightleftharpoons SE \rightarrow E + P \tag{6.1}$$

Here, S is the substrate, E the enzyme, SE a "complex" in which the molecules S and E are associated, and P the product of the reaction. After the complex SE had been formed by reversible association, indicated by the double arrow, internal processes in this complex transform S into the product, P, which separates from the enzyme, E. This second transformation requires a certain average time, which we call t_e. The inverse constant, $k_e = 1/t_e$ is called the "rate constant" of the reaction; it is the average number of substrate molecules a single enzyme molecule can transform in a second.

As the rate of supply of the substrate, S, increases, the enzyme molecules, E, released at the end of reaction (6.1) become reloaded again with the substrate more and more quickly. When the supply of S is very fast, all molecules of E are kept continuously occupied. Increasing still further the supply of S has now become useless. The enzymatic transformation, represented by the arrow from SE to $E + P$ in Eq. 6.1, has now become a bottleneck (in industrial parlance) or a rate-limiting reaction (in the terminology of physical chemistry); it now limits effectively the rate of the overall reaction. In photosynthesis, where S is produced by light, this means that we have passed from the light-limited state into the light-saturated, enzyme-limited state. It is as if soldiers were brought to port by railroad and then shipped overseas. The ships shuttle back and forth as fast as they can, but require certain time for the round trip. As long as trains arrive at a leisurely rate, ships have no difficulty in taking care of all arrivals; the whole transport operation is "train-limited." As the rate of train arrival increases, the whole operation passes from the "train-limited" to the "ship-limited" state. The essential point in the case of photosynthesis is that the number of "trains" can be increased at will (by increasing the intensity of light), while the number of "ships" is limited by the finite number of enzyme molecules in the cell.

One can surmise that the rate of the photochemical production of the substrate S in Eq. 6.1 remains proportional to the intensity of illumination, even when light saturation is reached; but that excess molecules of S, not taken care of at once by the enzyme, E, crowd the too-strongly illuminated cell, like unemployed workers crowd a labor exchange when there are not enough jobs. Two things can happen in this situation. The substrate molecules S may be *stable*. Then, after their supply had been stopped (for example, by the cessation of illumination), the accumulated supply will be worked up by the enzyme. In other words, photosynthetic production will continue for awhile in darkness. Alternatively, the light-produced substrate molecules S may be unstable. In this case, they will not hang around, but disappear, like soldiers drifting home when stranded in a port without enough ships to take them overseas. Experiments show that the production of O_2 and the reduction of CO_2 do *not* continue for a significant length of time after the cessation of illumination, however strong the latter had been. This suggests that the second alternative is correct; that is, that light produces *unstable* intermediates. These have to be stabilized by an enzymatic reaction; otherwise, they are lost by sliding back, or falling aside (*i.e.*, by back reaction or side reactions).

Figure 6.1 shows that the photochemical reaction imposes on the light curves of photosynthesis, $P = f(I)$, a slanting "roof":

$$P = k_i I \tag{6.2}$$

where k_i is a proportionality constant. The dark enzymatic reaction imposes on P a horizontal "ceiling":

$$P_{\max} = \frac{k_e[E_0]}{n} \tag{6.3}$$

where $[E_0]$ is the total concentration of enzyme molecules available in the cell, and n a small whole number (see below).

So far, we did not make any assumption about the mechanism of photochemical reactions, except for the natural one that its rate is proportional to the intensity of illumination. We now make use of Einstein's quantum theory of light, according to which light is absorbed by matter in discrete packages, the so-called energy *quanta* or *photons*. The energy content ϵ of a quantum is proportional to the frequency of the light, ν, and thus inversely proportional to its wavelength λ:

$$\epsilon = h\nu = \frac{hc}{\lambda} \tag{6.4}$$

where λ is expressed in cm; h is Planck's universal quantum constant ($h = 6.6 \times 10^{-27}$ erg sec), and c is the velocity of light, 3.0×10^{10} cm/sec).

Sixty years ago, Einstein formulated the basic law of photochemistry: *one* absorbed quantum causes the transformation of *one* molecule. This law is undoubtedly correct for the initial excitation of the absorbing molecule. But the ultimate result, measured in the number of substrate molecules transformed, or of the product molecules formed, may be quite different, depending on the efficiency of secondary reactions, which follow the primary excitation act. The number of molecules transformed by a single absorbed quantum is called the *quantum yield* of a photochemical reaction; its inverse, the number of light quanta needed to transform one substrate molecule (or to produce one product molecule), is called *quantum requirement*. If the quantum initiates a long *reaction chain*, the quantum yield may rise into the hundreds or thousands. If most primary reaction products are lost by back reactions, this yield may go down to small fractions of unity.

The proper way to measure the rate of photosynthesis in relation to light intensity is thus by the *number of molecules* transformed (that is CO_2 molecules consumed or O_2 molecules liberated) per *absorbed light quantum*. If Fig. 6.1 is drawn on this scale, with the abscissa representing the rate of absorption of quanta in einsteins/sec, and the ordinate the rate of liberation of O_2 (or consumption of CO_2) in moles/sec, the tangent to the curve (its slope) at its beginning is the *maximum quantum yield* of the process. This yield remains constant in the light-limited, linear part of the light curve, but declines as this curve bends towards the horizontal. The constant k_i in Eq. 6.2 is the maximum quantum yield of photosynthesis, as observed in the limiting case of weak light.

Equation 6.2 is based on Einstein's first law of photochemistry, whereas Eq. 6.3 is based on Guldberg and Waage's first law of reaction kinetics, the so-called *mass action law*. This law asserts that the rate of a chemical reaction is proportional to the concentration of the molecules participating in them. Thus, the rate of the second reaction in Eq. 6.1 is proportional to the concentration of the complex $[SE]$. (Concentrations are often designated by square brackets, $[A]$ meaning "concentration of A.") The coefficient k_e is the rate constant. The maximum possible value of $[SE]$ is $[E_0]$, the total concentration of the enzyme in the cell. The ceiling rate in Fig. 6.1 is defined by Eq. 6.3. The small whole number, n, is included in Eq. 6.3 because of the probability that not one, but several (perhaps 8) products P of reaction (6.1) must be formed for a single O_2 molecule to be liberated.

The above-suggested simple two-stage mechanism of photosynthesis explains the fundamental fact of light saturation. Mathematical analysis shows that it also explains the hyperbolical shape of some (although not all!) experimental light curves $[P = f(I)]$. It explains also the way in which *temperature* and certain *poisons* affect the rate of photosynthesis (see Fig. 6.1). Lowering the temperature, or adding certain poisons has no effect on the rate of absorption of light quanta, and thus also on the rate of supply of the substrate S; therefore, in the light-limited state, the rate of photosynthesis does *not* depend on these factors; but they do affect the ceiling rate, k_eE_0. Changing the temperature changes the rate constant, k_e, because the dark reactions generally go faster the higher the temperature. Adding enzyme poisons diminishes the number of available enzyme molecules, E_0, and thus reduces the ceiling rate $k_e[E_0]$. Many poisons, such as cyanide, act by combining with heavy metal atoms (for example, iron atoms) present in an enzyme molecule, and thus make the latter inactive. The proportion of such deactivated enzyme molecules increases with increasing concentration of the poison, until they are all immobilized, and the reaction rate is reduced to zero.

So far, so good; but more detailed, quantitative studies showed that the situation is more complex. Not one, but a whole series of enzymatic reactions, each with its own specific sensitivity to poisons and temperature changes, appear to be involved in the reaction sequence of photosynthesis. One of these enzymes may, under a given set of conditions, act as *the* rate limiting one, just like the narrowest bridge may limit the traffic-bearing capacity of a whole road. In the presence of certain specific poisons, another enzyme may become rate-limiting, as another bridge may become a bottleneck if it is under repair. Here is a striking example. The maximum rate of reduction of quinone to hydroquinone (or of ferricyanide to ferrocyanide) and of the liberation of oxygen by green cells or cell fragments in light (the so-called Hill reaction see Chapter 7) is about the same as that of photosynthesis. This suggests a common rate-limiting enzymatic reaction. And yet, upon addition of cyanide, the rate of photosynthesis goes down strongly, while that of the Hill reaction is not affected at all! This suggests that the common rate-limiting reaction is *not* sensitive to cyanide, but that the rate of some other enzymatic reaction, involved only in photosynthesis and not in the Hill reaction, is depressed by cyanide until this reaction becomes the bottleneck.

Biochemical analysis of the mechanism of CO_2-reduction in photosynthesis (see Chapter 17) led to the identification of more than a dozen enzymes involved in this process. The relation between these enzymes and the rate-limiting enzyme, whose presence had been derived many years earlier from the shape of the light curves (and other kinetic data), remains uncertain. Apparently, the main rate-limiting reaction in photosynthesis and in the Hill reaction is involved in the reaction sequence that is common to both processes, and not in the reduction of carbon dioxide, which does not occur in the Hill reaction.

PHOTOSYNTHESIS IN FLASHING LIGHT

If photosynthesis consists of a practically instantaneous primary photochemical reaction and an enzymatic dark reaction (or reactions), which require a certain average time, t_e, then the two reactions may be separated by the use of *flashing light.* Photochemical reactions offer to the experimentalist the convenient possibility of starting and stopping them as quickly and as often as desired by switching the light on and off. Also, one can easily produce light flashes, lasting only milliseconds or even microseconds, supplying a sufficient number of quanta to produce a measurable chemical change. In studying photosynthesis by means of flashing light, we want to send into a cell a practically instantaneous flash, containing enough photons to produce a measurable amount of oxygen. (An "instantaneous" flash can be defined here as one much shorter than the time, t_e, required for the completion of the rate-limiting enzymatic reaction.) Such flash illumination can be produced *mechanically,* by placing a slit in a rotating disc in the path of a strong, steady light beam, or *electrically,* by loading up a condenser and discharging it through a vacuum tube. (Recently, the use of lasers has been initiated to produce sufficiently intense monochromatic flashes lasting only nanoseconds.)

The two methods have their advantages and disadvantages. In the rotating disc technique, if a sufficiently intense light source is available, one can vary the flash energy, E_f (the total number of quanta supplied by the flash) within very wide limits, by making the slit broader or narrower, and by rotating the disc more or less rapidly. But if one goes down to flashes shorter than a millisecond, their energy becomes small. In the condenser-discharge technique, flash duration can be made very short—a few microseconds or even less—but unless one has available an unusually powerful condenser, the total energy of the flash, E_f, is not high—while one usually would like to make it strong enough to attain flash saturation (see below). Another limitation is set by the time needed to reload the condenser before it can fire again, which usually takes $\gg 0.1$ sec.

One does not usually need to measure oxygen yield from one flash. (Certain extremely sensitive methods, for example, observation of the phosphorescence of certain phosphors, which is suppressed by the slightest traces of oxygen, do permit, however, a rough estimate of this magnitude.) Usually, it is enough to measure the oxygen volume produced by a known number of repeated flashes. In the rotating slit technique, the frequency of flashes must be sufficiently small for the dark enzymatic reaction to be completed after each flash. Experiments showed that for this purpose, the dark periods in photosynthesis must last at least 0.1 sec, so that one has to operate with not more than ten flashes per second.

The first experiments on photosynthesis in flashing light were made by Robert Emerson and William Arnold in 1932. They exposed suspensions of Chlorella cells to condenser flashes lasting about 10^{-5} sec and measured the rate of oxygen evolution in relation to the energy of the flashes and the duration of dark intervals between them. They also observed the effects of temperature and of certain poisons on the oxygen production in flashing light. These experiments gave some unexpected results.

Emerson and Arnold found that if the energy of the flashes is progressively increased, the oxygen yield *per flash,* which at first grows propor-

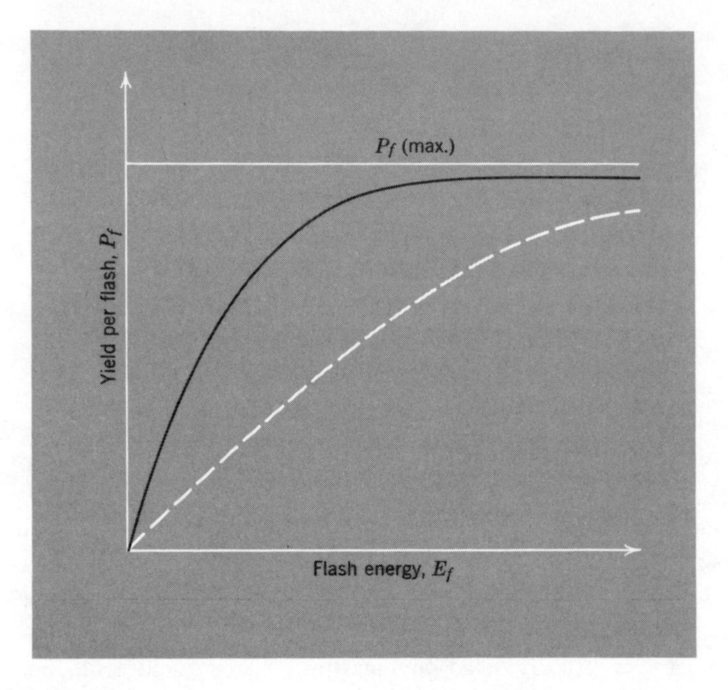

FIG. 6.2 Flash saturation of photosynthesis. Dashed line corresponds to shorter dark intervals or lower temperature. (R. Emerson and W. Arnold, 1932.)

tionately to this energy, finally shows saturation, approaching a maximum rate (Fig. 6.2)—just as the rate of photosynthesis in constant light. It must be noted that what is plotted in Fig. 6.2, in contrast to Fig. 6.1, is the *yield* P_f (per flash) and not the *rate* P (yield per unit of time exposure). The flash ceiling, P_f^{max}, is unexpectedly low. Before the relation between the enzymatic and photochemical stage in photosynthesis became clear, one did not expect saturation in flashing light to occur until each chlorophyll molecule had been given the chance to absorb a quantum of light in the flash (and thus to produce material to be worked over during the dark period). Instead, flash saturation was found to occur, in normal cells, already when only one out of 2500 chlorophyll molecules had received a quantum during the flash. We suggested above an interpretation of this observation, in terms of a limited number, $[E_0]$, of available enzyme molecules. Since enzyme molecules, and not chlorophyll molecules, have to "chew" on the photoproduct during the dark interval, it is their number and not that of chlorophyll molecules that determines the maximum yield obtainable from a flash. To explain Emerson and Arnold's results, one could thus suggest that one molecule of the limiting enzyme is available for 2500 molecules of chlorophyll. If several molecules of the photochemical intermediate (S in reaction 6.1) are supplied to each enzyme molecule in a single flash, one of them preempts it by forming the complex (SE) and all others go to waste. This consideration led to the concept of a *photosynthetic unit* of about 2500 chlorophyll molecules, associated with a single enzyme molecule (or, we would now say, with the entrance to a single enzymatic "conveyor belt"). This important concept was introduced by H. Gaffron and K. Wohl in 1936. Some structures, observed in electron-microscope pictures of chloroplast fragments, have been tentatively identified with photosynthetic units (see Chapter 8).

One correction needs to be introduced: we measure the yield per flash by the number of oxygen molecules produced; but the production of each oxygen molecule is likely to require not one, but several primary photochemical processes, and, therefore, put the enzymatic conveyor belt to work not once, but several times. How many? We shall see that the most likely figure is *eight* (two for each hydrogen atom transferred from H_2O to CO_2). The liberation, in a flash, of one oxygen molecule per 2500 chlorophyll molecules thus may mean that one enzymatic center is present per $\frac{2500}{8}$, that is, per about 300 chlorophyll molecules. The photosynthetic unit would then consist of three hundred, rather than of two and one-half thousand chlorophyll molecules.

Like green plants, photosynthesizing bacteria also contain photosynthetic units, that is, a large number of pigment molecules present per enzymatic center. However, this number is smaller than in green plants—about 50 instead of 300.

These numbers (300 and 50) are typical of many normal, healthy cells, but may vary widely from plant to plant. For example, they are much lower for chlorophyll-deficient variegated leaves.

We have thus interpreted the maximum flash yield of photosynthesis as measure of the available amount of a rate-limiting enzyme, and estimated that the ratio between the concentrations of chlorophyll and of this enzyme is, in typical healthy plant cells, about 300 to 1.

The need for something like the postulated photosynthetic unit can be easily seen a priori. In order to absorb enough light, a cell must contain as much as 0.1 mole chlorophyll per liter. If each chlorophyll molecule were provided with its own assortment of enzyme molecules, the latter would have to be present in the same concentration. However, enzymes are protein molecules, with a molecular weight of 10^5–10^6, while chlorophyll has a molecular weight of about 10^3. Each enzyme molecule, therefore, requires 100–1000 times the space of a single chlorophyll molecule. There is not enough space in the cell for so many enzyme molecules! Thus, many chlorophyll molecules simply *have* to share a single enzyme molecule. Fortunately, the effectiveness of enzymes is great enough for this sharing. In direct sunlight, a chlorophyll molecule will absorb photons at the rate of one to ten per second, while a good enzyme can easily transform a thousand or ten thousand substrate molecules each second; it can thus keep pace with the substrate supply from several hundred chlorophyll molecules.

The flashing light experiments can be used for another determination: that of the "working time," t_e, of the rate-limiting enzyme. For this purpose, we measure the yield of oxygen per flash as function of the dark period between flashes, t_d. (In this experiment, care must be taken to use sufficiently strong "saturating" flashes.) As the duration of the dark intervals increases, the yield per flash increases too, until the intervals reach the order of 0.1 second. Then the curve approaches saturation. This suggests that the rate-limiting enzyme requires about 0.1 sec to work up *practically all* the supply of substrate it had received during the flash. The *average* time, t_e, required to transform a substrate molecule is smaller—about 0.02 sec at room temperature.

Changes in temperature were found by Emerson to affect not the maximum yield per flash, but the dark interval needed to obtain it. This, too, agrees with the interpretation of the maximum yield as a measure of the available amount of a limiting enzyme. This conclusion is further confirmed by the effect of the addition of potassium cyanide on the flash yield; thus poison, too, does not change the maximum flash yield, but lengthens the required dark interval.

We can thus derive, from flashing light experiments, *two* independent constants:

$$[E_0] = \frac{\text{Chl}_0 n}{2500}$$

where Chl_0 is the total amount of chlorophyll; and

$$k_e = \frac{1}{t_e} = \frac{1}{0.02} = 50\ \text{sec}^{-1}\ (\text{at } 20°\text{C})$$

Now comes a very satisfying calculation: if the enzyme that limits the steady rate of photosynthesis in *constant* light is the same one that limits the yield *per flash* in *flashing* light, then the saturation rate of photosynthesis (P^{max}) in constant light (Fig. 6.1) must be equal to $k_e[E_0]/n$ (Eq. 6.3). This, in turn, should be equal to the product of the two above-determined constants, or about 0.02 Chl_0 per second. This means that the maximum rate of photosynthesis *in constant light* should be one oxygen molecule each 50 seconds per molecule of chlorophyll. And this is what it actually is! It has been known since the pioneer measurements of Richard Willstätter and Albert Stoll (1913–1918) that healthy, fully-active leaves, abundantly supplied with carbon dioxide and light, can produce one molecule of O_2 (and consume one CO_2 molecule) every 20–30 seconds per chlorophyll molecule present in them. Willstätter called this the "assimilation time" of the leaves (carbon dioxide assimilation being, we recall, another term for photosynthesis). For Chlorella cells, somewhat higher assimilation times, 40 or 50 seconds, have been found. This is close enough to the value calculated from flashing light experiments to assert that the latter actually permit factorization of the maximum rate in constant light, $k_e[E_0]$ into the two factors k_e and $[E_0]$.

However, as always in more precise study of biological phenomena, this simple relation proved to be *too* simple. Photosynthesis involves not a single one, but a number of enzymatic reactions, and more than one of them can affect the maximum rate, in flashing as well as in constant light. Subsequent flashing light experiments by James Franck and S. Weller in the United States, by Hiroshi Tamiya and co-workers in Japan, and by Bessel Kok in Holland, gave evidence of such complications. According to Tamiya, if the flash lasts several *milli*seconds (instead of *micro*seconds, as in Emerson's experiments), the maximum yield per flash rises above the saturation plateau in Fig. 6.2, and becomes dependent on temperature—despite the fact that a few milliseconds is still a short period compared to the above-calculated "working time" (about 10 milliseconds) of the "Emerson-Arnold enzyme." Kok explained this behavior by postulating a more complex mechanism, involving two successive enzymatic steps. The maximum yield of a longer flash measures, according to Kok, the combined "reservoirs" of both enzymes $[E_0' + E_0'']$, rather than that of one enzyme only, as does the maximum yield of a shorter flash. The observed working time is, according to Kok, a function of the working times of both enzymes, more closely related to that of the second than to that of the first one.

One final cautionary remark: When there are two or more narrow bridges on a road, the maximum traffic it can bear is affected by all, and not only by the narrowest one. In the same way, if several enzymatic "bottlenecks" exist in a sequential reaction, all of them (and not only the narrowest one) affect the saturation rate of the overall process. For example, if two reactions in a series have the same maximum rate, V max, the maximum rate of the overall process will be (under certain conditions) only V max/2. The maximum rate of the overall reaction can be equated with the maximum rate of a single "limiting" step only if the limits imposed by all other steps lie far above that of the "limiting" step.

Despite these complications, it seems certain that the ratio 300 to 1 represents a significant relation between the number of pigment molecules and the number of enzymatic centers present in typical healthy green cells.

Ultimately, *kinetic* data will have to be brought into line with *biochemical* data, that is, with the amounts and action times of specific enzymes known to take part in photosynthesis (see Chapter 17). We are as yet far from achieving this aim.

Chapter 7

Taking Photosynthesis Apart. II. Photochemical Activities of Chloroplasts and Chlorophyll Solutions

CHLOROPLASTS: THE HILL REACTION[1]

We will now describe another, more direct way to separate photosynthesis into stages. From kinetic evidence (that is, rate measurements), we turn to attempts to break the cells, fractionate the cell material, and study the photochemical behavior of these fractions. This approach has long seemed hopeless. The alternative seemed to be: either live cells, capable of complete photosynthesis or dead, photochemically inert cell debris! The situation began to change thirty years ago. In 1937, an English plant biochemist, Robert Hill, following up an old observation that dry leaf powders exposed to light liberate a small amount of oxygen, discovered that this short-lived effect can be prolonged by supplying suspended leaf material with certain iron salts, such as ferric oxalate. This became known as the Hill reaction, and proved to be a very significant discovery, with which we must deal in some detail.

In Chapter 8, we will see that cells capable of photosynthesis contain,

[1] See Chapter 16 for Hill reaction studies in fractions of chloroplast material supposedly enriched in one of the two photochemical systems.

with very few exceptions, small pigmented bodies called *chloroplasts*. Microscopic observations with luminescent and motile bacteria, which are attracted by the slightest traces of oxygen, had shown long ago that chloroplasts are the sites of oxygen evolution in photosynthesis. This, and the localization of pigments in chloroplasts, suggested that they are the photosynthetic organelles of plants. When leaves are minced mechanically, for example, by means of a blender, or a pestle in a mortar, the resulting mash can be separated by means of a centrifuge into fractions, some of which contain, mostly or exclusively, whole or broken chloroplasts. These fractions prove to be the ones capable of carrying out the Hill reaction. Spinach leaves have been used more often than any others for this purpose, partly because fresh spinach is easily obtainable on the market.

Methods for preparing photochemically active suspensions from fresh, healthy leaves were improved by many investigators who rightly saw in Hill's finding a first hopeful glance into the black box of photosynthesis. It suggested that one part of photosynthesis, the liberation of oxygen in light, could be reproduced in an extracellular preparation—a suspension contaning chloroplasts or chloroplast fragments, but no cytoplasm, nuclei, or mitochondria.

It was soon found that ferric oxalate is not the only chemical to produce sustained photochemical oxygen production from chloroplast suspension. Hill himself observed that ferricyanide is even better for this purpose. Otto Warburg, in Germany, noted that quinone (orthobenzoquinone) is quite active. Subsequent studies have shown that a large number of compounds of the quinone type, as well as many organic dyes, with a similar, "quinonoid," structure, have the same capacity. All these compounds have one property in common: they are good oxidants. Ferric salts, containing the ion Fe^{3+}, can be reduced to ferrous salts, containing the ion Fe^{2+}, by the addition of an electron; quinones, Q, can be reduced to hydroquinones, QH_2, and dyes, D, to colorless leucodyes, DH_2, by addition of two hydrogen atoms:

$$Fe^{3+} + e^- \rightarrow Fe^{2+}$$
$$Q + 2H \rightarrow QH_2, \text{ or } D + 2H \rightarrow DH_2$$

It thus appeared tempting to interpret the Hill reaction as "photosynthesis with a substitute oxidant." Referring to Fig. 5.4, the CO_2-reducing enzymatic system, represented by the *upper* horizontal arrow, seems to be lost (or damaged) in the preparation of the chloroplast suspension, so that another oxidant (hydrogen acceptor), which may be a ferric salt, a quinone, or a dye, must be substituted for the system CO_2/CH_2O.

Quantitative studies of the Hill reaction confirmed that it was in fact a photochemical oxidation of water. About the same maximum quantum efficiency and the same maximum rate of oxygen evolution (for a given amount of chlorophyll) were found in the Hill reaction as in photosynthesis.

The Hill reaction thus represents the long sought-after residual activity of a fraction of the complete photosynthetic apparatus left intact after mechanical destruction of the cells and isolation of chloroplast material.

These findings revived the old controversy as to whether the photochemical reaction of photosynthesis is concerned primarily with carbon dioxide or with water. Since the earliest investigation, it was known that H_2O and CO_2 are the two reactants involved in photosynthesis. Two attitudes had emerged: plant physiologists, botanists, and biochemists were concerned, above all, with the conversion of CO_2 into organic nutrients; photosynthesis appeared to them as "reduction of carbon dioxide to carbohydrates" in light. This attitude found its extreme expression in dividing photosynthesis into the following two steps (see Eq. 1.5):

$$CO_2 \xrightarrow{\text{light}} C + O_2; \; C + H_2O \rightarrow (CH_2O) \tag{7.1}$$

However, some chemists (among them G. Bredig in 1914) suggested that the light reaction of photosynthesis may be primarily concerned with water; for example:

$$2H_2O \xrightarrow{\text{light}} O_2 + 4H; \; 4H + CO_2 \rightarrow (CH_2O) + H_2O \tag{7.2}$$

but this view found little attention at that time.

As stated in Chapter 5, we now realize that photosynthesis is neither "decomposition of carbon dioxide" as in Eq. 7.1, nor "decomposition of water," as in Eq. 7.2, but an *oxidation-reduction reaction* between H_2O and CO_2, an uphill transfer of four hydrogen atoms from H_2O to CO_2. The two reactants are equally important, and play a symmetric role in the overall process.

However, a legitimate question remains whether, in Fig. 5.4, the photochemical, energy-storing step is closer to the removal of hydrogen from water, or to the addition of hydrogen to carbon dioxide. It is like asking whether, on an uphill waterway, the locks are located at the lower or at the upper end. Considerations of comparative physiology, especially by C. B. van Niel (see Eq. 2.2), and the discovery of the Hill reaction, suggested that the primary photochemical reaction is more closely associated with the *dehydrogenation* of water than with the *hydrogenation* of carbon dioxide. It is, however, still incorrect to say, as it is often done, that the primary process in photosynthesis is "photolysis of water," which would suggest dissociation of H_2O into the molecules H_2 and O_2, or into the radicals OH and H.

One experimental finding, disproving the hypothesis in Eq. 7.1, is the demonstration that the source of O_2 in photosynthesis is the water molecule (as in Eq. 7.2) and not CO_2. This proof was given by Sam Ruben and Martin Kamen in Berkeley in 1941. They found that when plants were supplied ^{18}O-enriched water, the isotopic composition of the evolved oxygen was that of water and not that of CO_2.

The reducing capacity of the chloroplasts in the Hill reaction seemed at first to be insufficient to reduce CO_2 and thus to carry out complete photosynthesis; only oxidants with redox potentials more positive than 0.0 volt could be easily utilized for this reaction. More recent experiments suggested, however, that if back reactions can be avoided, even more reluctant oxidants can be reduced. In fact, it has been found possible to reduce, by means of illuminated chloroplasts, the important biological catalyst $NADP^+$ (oxidized nicotinamide adenine dinucleotide phosphate), with a potential of about —0.35 volt, if certain catalytic components (present in green plants and believed to be involved in photosynthesis) are added. Such observations were first made by W. Vishniac and S. Ochoa in New York in 1951 and confirmed by Anthony San Pietro (then at the Johns Hopkins University) and Daniel Arnon (at Berkeley). The necessary catalytic compounds include an iron-containing protein called ferredoxin and an enzyme called ferredoxin-NADP-reductase; these will be discussed in Chapter 17.

It is now widely assumed (although still not quite certain) that in true photosynthesis, the photochemical stage ends in the reduction of $NADP^+$ to NADPH (and production of a certain amount of so-called high energy phosphate, adenosine triphosphate, or ATP; see Chapters 17 and 18). Therefore, there seems to be no reason why chloroplasts, provided with catalytic supplements, should not also bring about the reduction of CO_2. Arnon and co-workers at Berkeley found, in fact, that illuminated chloroplast suspensions, provided with a proper assortment of catalytic "co-factors," can transfer ^{14}C from labeled carbonate to organic compounds previously found to occur as intermediates in the reduction of CO_2 in live cells (see Chapter 17). This suggests that the true reducing power of illuminated isolated chloroplasts is not lower than that of the same chloroplasts in whole cells.

In Arnon's earlier experiments, the rate of CO_2 reduction (and O_2 liberation) with illuminated chloroplast preparations appeared quite low compared to that observed in photosynthesis of whole cells containing the same amount of chlorophyll. Recently, however, D. Walker in England was able to raise the rate to 10–15% of that of intact leaves; and in 1967, R. G. Jensen and J. Bassham in Berkeley obtained, *in strong light,* a rate of oxygen liberation equal to about 60% of that of intact leaves, but only in the presence of CO_2 of much higher concentration than in the air; this rate could be maintained for 6–10 minutes. The decisive factors in these experiments seem to be (1) the use of a suspension medium (developed by Norman Good of Michigan University) containing high-molecular-weight compounds, which seem to preserve intact the chloroplast membranes and prevent rapid loss of certain enzymes; and (2) quick separation of the chloroplasts from the rest of the cell material. Recently, we have found that the quantum yield of O_2 evolution by such chloroplast suspensions *in weak light* may reach 30–40% of that of intact Chlorella cells under identical conditions.

To sum up, the upper horizontal arrow in Fig. 5.4 can be reconstructed in chloroplast suspension; but as yet, such reconstructed systems have been able to operate only with reduced efficiency, and not for long.

Why live cells devote themselves to the reduction of carbon dioxide and "accept no substitutes," is one of the great wonders of photosynthesis. The rule is however, not without exceptions. For example, if *Chlorella* cells are placed in a solution of benzoquinone, they begin to act like chloroplast suspensions, that is, to liberate oxygen in light, and reduce quinone, but leave carbon dioxide alone. These cells are dead; they do not respire and cannot be revived by washing-out the quinone. Other Hill oxidants cannot be used in the same way. In many cases, they simply do not penetrate into healthy cells (although their penetration can be forced e.g., by washing the cells with glutaraldehyde.)

One well-known strong oxidant that does easily penetrate into cells is free oxygen, and the striking property of the photosynthetic apparatus is its refusal to substitute O_2 for CO_2 as hydrogen acceptor. Substituting O_2 for CO_2 would mean photosynthesis running in a cycle:

$$H_2O + \frac{1}{2}O_2 \xrightarrow{\text{light}} \frac{1}{2}O_2 + H_2O \tag{7.3}$$

Certain observations suggest that such a "short-circuiting" of photosynthesis can in fact occur, but only under special conditions.

Chloroplasts, freed from all adhering mitochondria, do not respire (that is, do not use up oxygen in the dark). This suggests a neat division of the two functions, photosynthesis and respiration, between the chloroplasts and the mitochondria.

Simultaneous occurrence of respiration and photosynthesis adds a difficulty to all rate measurements of photosynthesis in live cells because one process undoes what the other does. In light, one can measure only their difference, while respiration can be measured by itself in darkness. To calculate the rate of photosynthesis, one has to postulate that respiration remains the same in light. This assumption can be tested by tracer experiments in which heavy oxygen isotope is used. Such experiments, by Allan Brown and co-workers at the University of Minnesota, showed the rate of O_2 uptake in dark and in light to be not too different, at least, in chloroplast-bearing algal cells. (However, additional oxygen uptake in light, so-called *photorespiration,* does occur in many plants in certain spectral regions.)

In chloroplast-bearing plants, one can say that photosynthesis takes place in chloroplasts and respiration elsewhere in the cell. Of course, this spatial separation cannot be absolute, since the products of photosynthesis, such as the carbohydrates, sooner or later diffuse into the cytoplasm, to be used there as substrates of respiration. Inversely, not only carbon dioxide produced by respiration, but probably also certain respiration intermediates can diffuse into the chloroplasts, and interfere there with the enzymatic processes of photosynthesis. In the primitive, chloroplast-free blue-green algae, interaction between photosynthesis and respiration is much livelier than in the chloroplast-bearing higher plants; at least, Allan Brown found a marked effect of light on O_2-uptake by these algae. In the course of evolution, plants must have found it advantageous to separate respiration from photosynthesis; this may have permitted them to keep all life processes, supported by respiration, on an even keel, day or night, rain or shine.

PHOTOCHEMISTRY OF CHLOROPHYLL SOLUTIONS

The green pigment chlorophyll, (Chl *a*), is a common constituent of all photosynthesizing plant cells. In addition, these cells contain also a variable assortment of other pigments (see Chapters 8 and 9). Chlorophyll *a* could be a physical agent in photosynthesis, collecting light energy and making it somehow available for photosynthesis, or it could serve as a photocatalyst; that is, act as a light-activated chemical catalyst that takes an active, albeit reversible, part in the photosynthetic reaction. This question encourages studies of the photochemical properties of chlorophyll (and other plant pigments) outside the plant cell. Nonphotochemical oxidation-reduction catalysts, such as the iron-containing proteins called cytochromes, operate by reversible participation in the transfer of electrons (or hydrogen atoms). If the transfer of an H-atom (or an electron) from one compound (the reductant) to another compound (the oxidant) does not easily occur by itself (although it is permitted by the order of redox potentials), the provision of an intermediate system will often facilitate it. The intermediate oxidant takes the H-atoms or electrons from the reductant and gives them up to the oxidant. It thus serves as a go-between in an otherwise difficult chemical transaction. A *photocatalyst* may act similarly, except that it requires excitation by light to ply its trade. If the end result of a reaction is increase in free energy (that is, if the oxidation-reduction occurs against the gradient of redox potentials, as in photosynthesis), excitation of the catalyst is indispensable to make the reaction possible.

We are thus interested in knowing whether chlorophyll *in vitro* can serve as an oxidation-reduction photocatalyst, and whether it can mediate oxidation-reduction reactions involving storage of light energy.

Many dyes are easily reduced to colorless leucodyes and reoxidized back to colored dyes. A well-known example is methylene blue ($E_0{'}$ = about 0.0 volt), which is often used in the reconstruction of biological processes; it is easily reduced to colorless leuco-methylene blue. This dye, as well as its close relative, the violet dye thionine, can bring about, in light, the oxidation of ferrous ions:

$$\text{dye} + 2\text{Fe}^{++} \underset{\text{dark}}{\overset{\text{light}}{\rightleftharpoons}} \text{leucodye} + 2\text{Fe}^{+++} \qquad (7.4)$$

If one strongly illuminates a solution containing the dye and a ferrous salt, the blue (or violet) solution is decolorized within a few seconds. It quickly regains its color in darkness, showing that the bleached system had a higher free energy than the colored one. It is about the best imitation we know of the postulated primary photochemical process of photosynthesis. Light is utilized in this experiment, as in photosynthesis, for an uphill oxidation-reduction reaction against the gradient of electrochemical potential. In contrast to photosynthesis, however, no enzymatic agents are present to stabilize the energy-rich products, and the stored light energy is rapidly dissipated by back reaction between ferric ions and leuco thionine (or leuco-methylene blue). In our laboratory, reaction 7.4 was carried out in an emulsion of ether in water. The leucodye, formed in light, was extracted into ether, while the ferric ions stayed in water. The two solutions, the aqueous and the ethereal, could be separated in a separatory funnel. The products were thus prevented from reacting back. When the two solutions were stirred together again in the presence of alcohol, which makes them mutually soluble, the delayed back reaction took place and the color returned. (This is analogous to the products of photosynthesis, sugar and oxygen, reacting back in respiration.)

E. Rabinowitch and J. Weiss, in 1937, searched for a similar behavior of chlorophyll solutions in *methanol,* but found that these solutions can be reversibly *oxidized* (by ferric ions), rather than reversibly *reduced* (by ferrous ions).

$$\text{Chl (in methanol)} + Fe^{+++} \underset{\text{dark}}{\overset{\text{light}}{\rightleftharpoons}} Chl^{+} + Fe^{++} \qquad (7.5)$$

In 1948 the Russian physical chemist, A. A. Krasnovsky, found that chlorophyll *a,* dissolved in *pyridine,* can be reversibly reduced in light by ascorbic acid ($E_0' = 0.0$ volt), to a pink "eosinophyll."

$$\text{chlorophyll (in pyridine)} + \text{ascorbic acid} \underset{\text{dark}}{\overset{\text{light}}{\rightleftharpoons}} \text{eosinophyll} + \text{dehydroascorbic acid}$$

The basic nature of the solvent (pyridine) seems to be essential for the display of oxidizing, instead of reducing properties of chlorophyll. In the dark, Krasnovsky's reaction goes back, showing that it is associated with storage of free energy. Krasnovsky and co-workers went one step further, showing that reduced chlorophyll can be reoxidized by a variety of oxidizing compounds. The net result is photocatalyzed reduction of these compounds by ascorbic acid. In many cases, this reaction leads to a net storage of chemical energy. For example, the well-known biological catalyst, riboflavin ($E_0' = -0.2$ volt) is reduced by ascorbic acid ($E_0' = 0.0$ volt) in an illuminated chlorophyll solution, overcoming an adverse potential gradient of 0.2 volt.

Krasnovsky suggested that light-excited chlorophyll molecules are reduced, in vivo, at the cost of water, as they are in vitro at the cost of ascorbic acid, and restored by reducing an appropriate intermediate, which in turn reduces carbon dioxide. However, Krasnovsky's experiments showed only reduction by ascorbic acid ($E_0' = 0.0$ volt) of compounds with redox potentials down to —0.2 volt, thus bridging one sixth of the 1.2 volt gap that has to be bridged in photosynthesis, where H_2O serves as reductant.

We will see in later chapters that photosynthesis probably involves two sets of primary photochemical reactions. One of them may involve photoreduction of an organic substrate by one pigment system, and the other, photooxidation of water by another pigment system (see Chapters 13–16). If Chl *a* is the photocatalyst, it will be oxidized in the first reaction and reduced in the second. A reaction between the oxidized form of chlorophyll *a* produced in one reaction and the reduced form produced in the second reaction is needed to close the sequence and restore the photocatalytic system to its original state. One is tempted to recall in this connection the above-mentioned tendency of chlorophyll for reversible photooxidation in alcoholic solution, and reversible photoreduction in pyridine solution. A remote, but plausible, analogy may exist between these two reactions in vitro and the two reactions of Chl *a* (probably located in different molecular environments) in photosynthesis. This is only a speculation, but it does invite systematic experiments.

Finally, we shall see in Chapter 14 that only a small fraction of all chlorophyll *a* molecules (perhaps only one per photosynthetic unit) is likely to engage in photocatalytic activity, while all others serve, similarly to accessory pigments, merely as "physical" excitation energy suppliers.

Chapter 8

Structure and Composition of the Photosynthetic Apparatus

CHLOROPLASTS, GRANA, LAMELLAE

It was said before that photosynthesis requires a certain subcellular structure. Probably, it is needed to direct the reaction sequence, and to prevent unstable intermediates from getting off their tracks.

Not long ago, chemical processes inside a living cell were considered as reactions in a homogenous solution confined in a bag. True, certain structures, granules or fibrils, had been early observed in the cell under the light microscope, but these were skeptically considered as artifacts resulting from slicing or staining. Subcellular particles clearly visible under the microscope, such as cell nuclei or chloroplasts, were considered as "bags within bags."

The invention of the electron microscope, with a much stronger resolving power (roughly speaking, 10^{-7} instead of 10^{-5} cm), brought unequivocal proof that the nuclei, the chloroplasts, and other "organelles," imbedded in the cytoplasm, as well as the cytoplasm itself, possess an internal structure undoubtedly significant for their metabolic functions.

The chloroplasts of the higher plants are shown in Fig. 8.1 as they

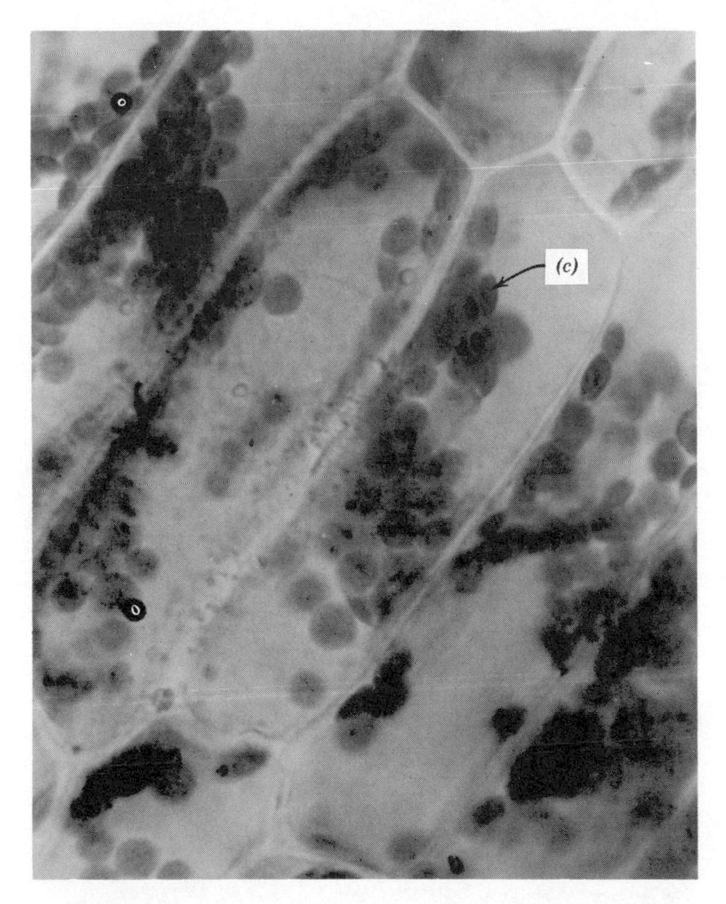

FIG. 8.1 Chloroplasts (*c*) in leaf cells (optical micrograph). (Courtesy of D. Paolillo.)

appear under the light microscope. They are more or less ellipsoidal bodies, 5–10μ in diameter, or about one fifth or one tenth the size of the cell (one μ is a thousandth of a millimeter, or 10^{-4} cm). In algae, the shapes of chloroplasts are more varied. In elongated algal fronds, they may be long, twisted bands. In some approximately spherical cell, they are single star-shaped bodies. The unicellular green alga Chlorella, often mentioned in this text, contains a single chloroplast shaped like a bell, adhering to the inside of the cell wall (Fig. 8.2).

The first structures discovered within chloroplasts by means of the

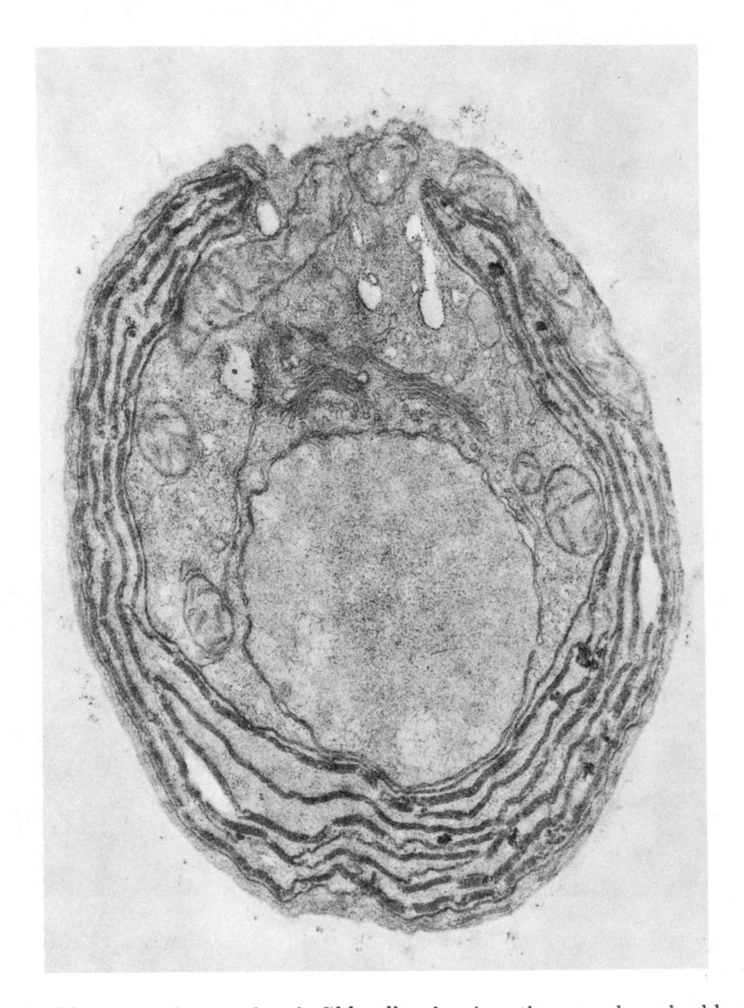

FIG. 8.2 Electron micrograph of *Chlorella* showing the cup-shaped chloroplast, magnification 24,000×. (Courtesy of T. Bisalputra, University of British Columbia, Canada.)

electron microscope were the so-called *grana,* flat tablets about 0.5 μ in diameter and about 0.3 μ thick. Several dozen grana may be found in a single chloroplast (Fig. 8.3).

Later, it became clear that in *intact* chloroplasts, the grana are parts of a system of *lamellae,* stretching through the chloroplast (Fig. 8.4).

Some chloroplasts seem to contain only lamellae and no grana at all, or only vague outlines of them, as in Fig. 8.5.

Figure 8.6 shows a slice through a leaf of corn (*Zea mays*), in which the chloroplasts in the leaf parenchyma are clearly granular, while those in the vascular tissue show only lamellae. It seems that grana are approximately cylindrical regions formed by stacking of electron-dense sections of the lamellae.

The lamellar structure is found not only in the ellipsoidal chloroplasts of the higher plants, but also in algal chloroplasts of other shapes, although in the latter, the lamellae may be involuted rather than plane-

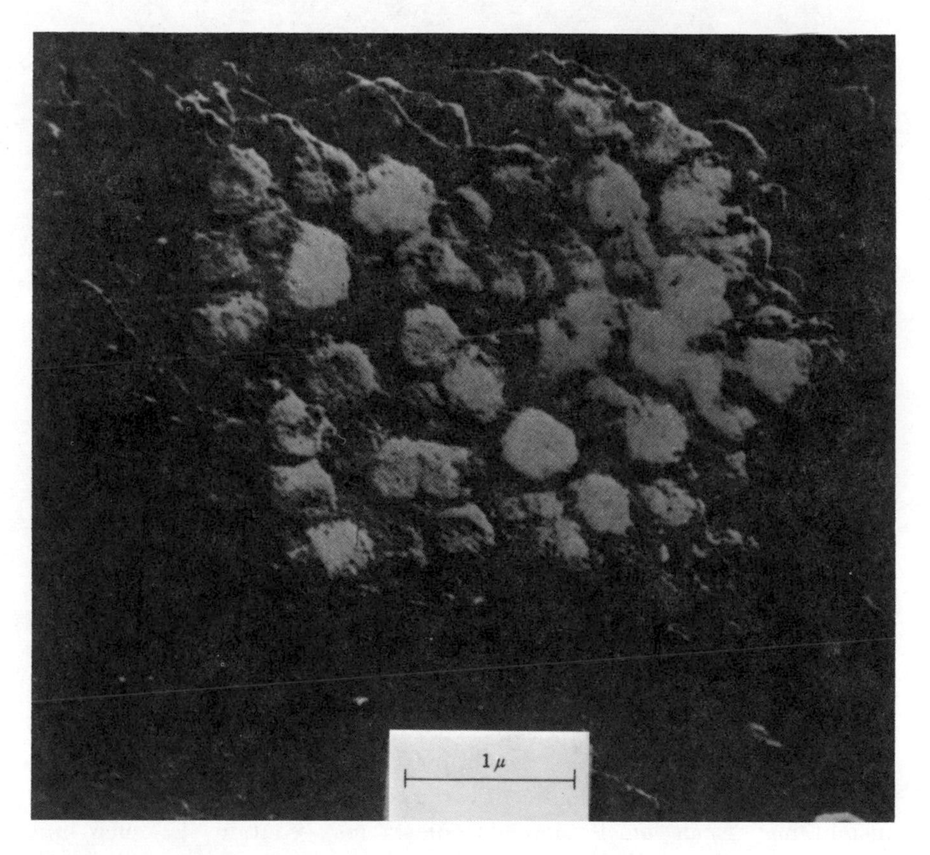

FIG. 8.3 Grana in a collapsed chloroplast of maize (*Zea mays*). (A. E. Vatter, 1952.)

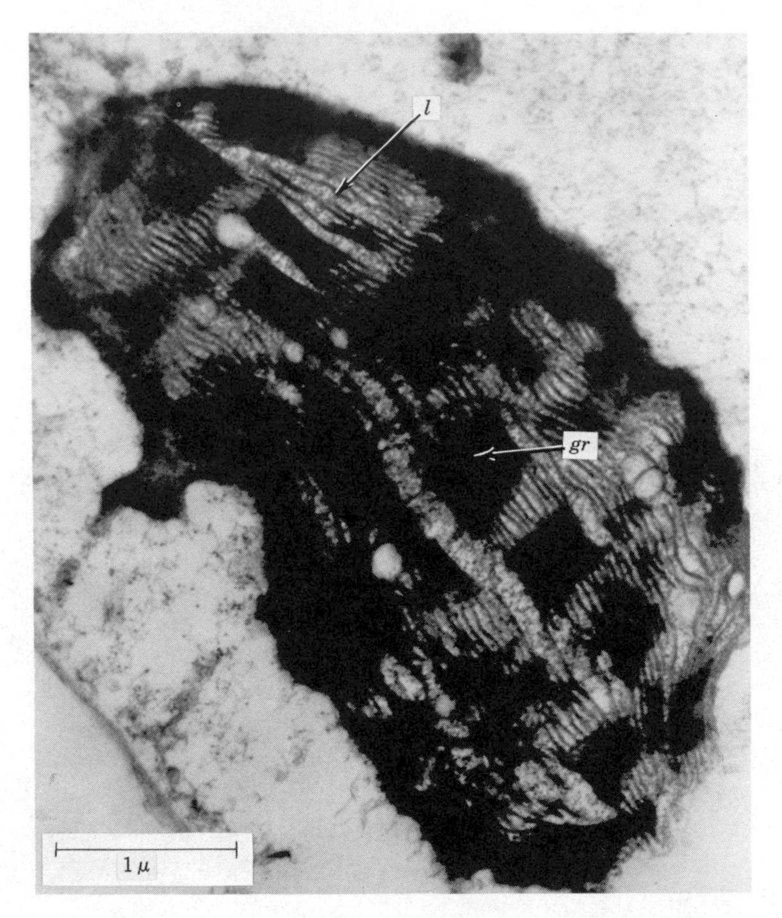

FIG. 8.4 Lamellae inside and outside the grana in maize (*Zea mays*). (*gr,* grana; *l,* lamellae.) (A. E. Vatter, 1955).

parallel (Figs. 8.2 and 8.7). The most primitive of plants, the blue-green algae (*Cyanophyceae*), contain no separate chloroplasts at all; but they do contain lamellae, extending through the cytoplasm.

The lamellae are a special case of two-dimensional cellular structures called membranes. Biological membranes generally consist of two superimposed layers; such a double layer is often referred to as the "unit membrane." Each lamella in chloroplasts may contain *two* such unit membranes, one on top of the other. W. Menke (Köln, Germany) has observed in photosynthetic bacteria more or less spherical "sacks" with

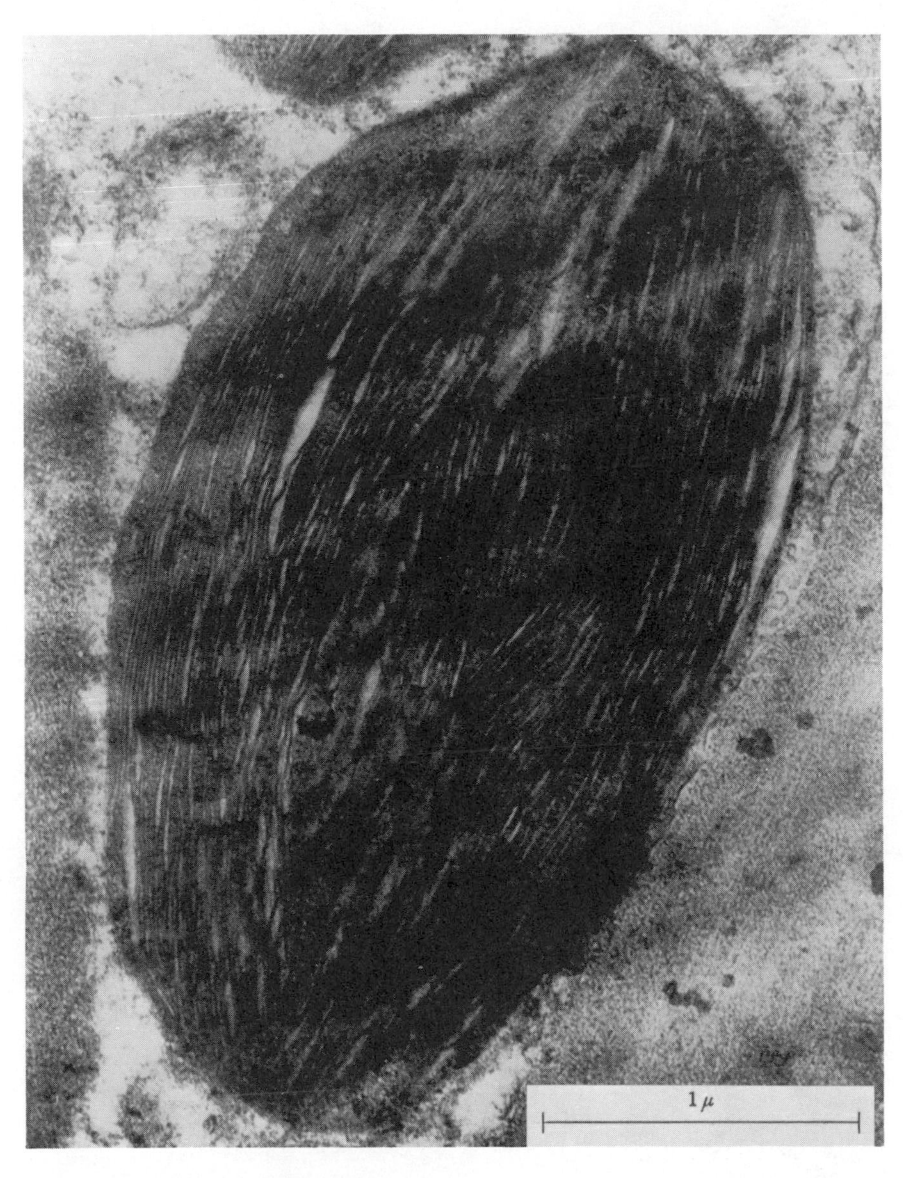

FIG. 8.5 Chloroplast from a light green part of the inner leaf of a young shoot of *Aspidistra elatior*. Lamellae without grana. (H. Leyon, 1954.)

double-layered walls, which he called *thylakoids* (Fig. 8.8). (These thylakoids are *not* "rooms without doors and windows"; extensive interconnections exist between them). The lamellae observed in higher plants may be flattened thylakoids, or assemblies of such thylakoids, fused together into flat "double pancakes."

With improved techniques of electron microscopy, more details were observed in chloroplast pictures, and more elaborate structures were suggested for the lamellae. We cannot enter here into the description of these details, particularly since most of them still remain speculative.[1] Instead, we return to the "photosynthetic units," whose existence has been derived in Chapter 6 from the kinetics of photosynthesis.

[1] See articles by W. Menke, and by R. B. Park and D. Branton, and by T. E. Weier, C. R. Stocking, and L. K. Shumway, in *Energy Conversion by the Photosynthetic Apparatus,* Brookhaven National Laboratory, January, 1967.

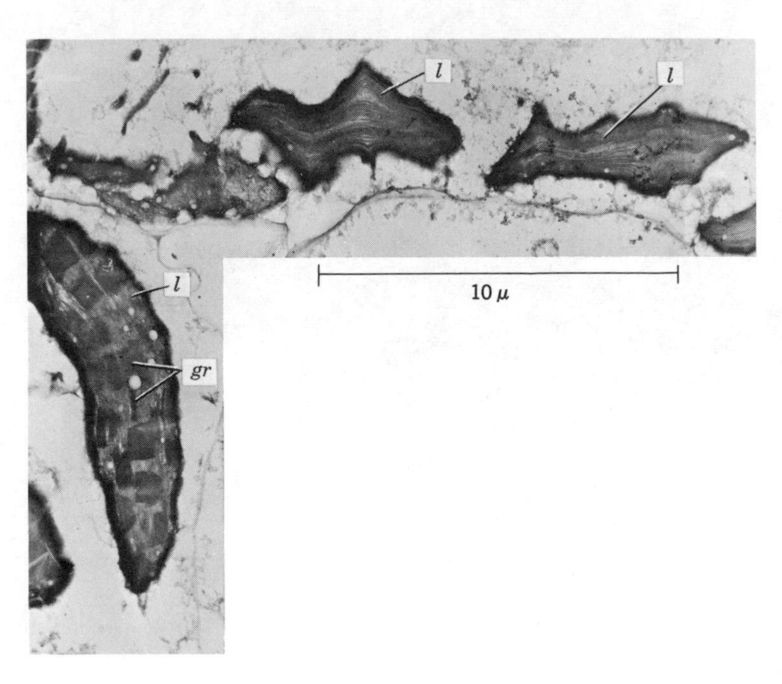

FIG. 8.6 *Zea mays* leaf section, showing granular chloroplasts in the leaf parenchyma (left) and nongranular chloroplasts in the vascular bundle (top right). (*gr*. grana; *l,* lamellae.) (A. E. Vatter, 1955.)

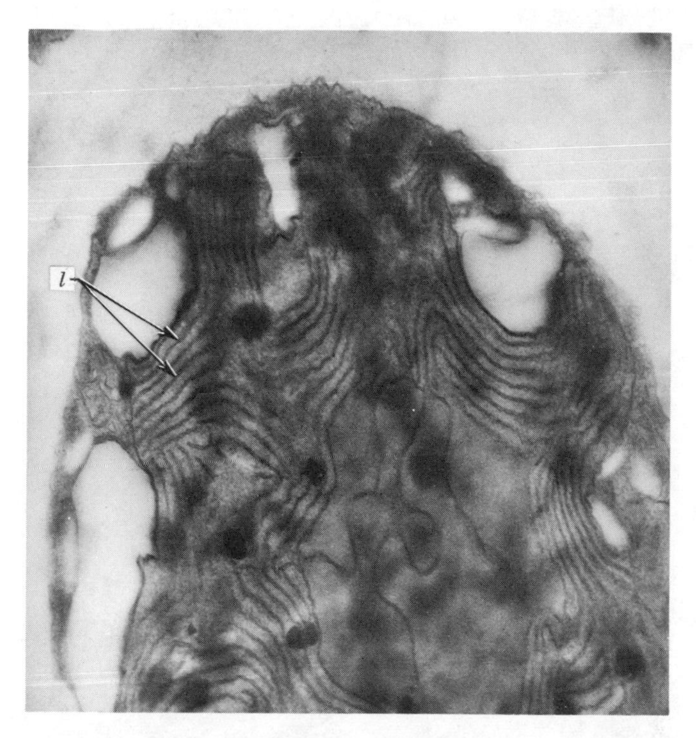

FIG. 8.7 Lamellae (*l*) in red alga *Porphyridium cruentum* (Marcia Brody and A. E. Vatter, 1958.)

PHOTOSYNTHETIC UNITS

In Chapter 6, it was noted that photosynthesis requires a large number of light-absorbing pigment molecules to supply excitation energy to a much smaller number of enzymatic "centers." Experiments were described there that suggested cooperation of about 300 chlorophyll *a* molecules with a single enzymatic center. These assemblies were called "photosynthetic units." Certain granular structures in chloroplast lamellae, observed under the electron microscope can be tentatively identified with these kinetic units. A "cobblestone pavement" picture of chloroplast lamellae was first observed in 1952 by E. Steinmann in Zurich. It appeared much clearer in pictures obtained by R. B. Park and J. Biggins at Berkeley, California, in 1961 (Fig. 8.9). The structure is particularly

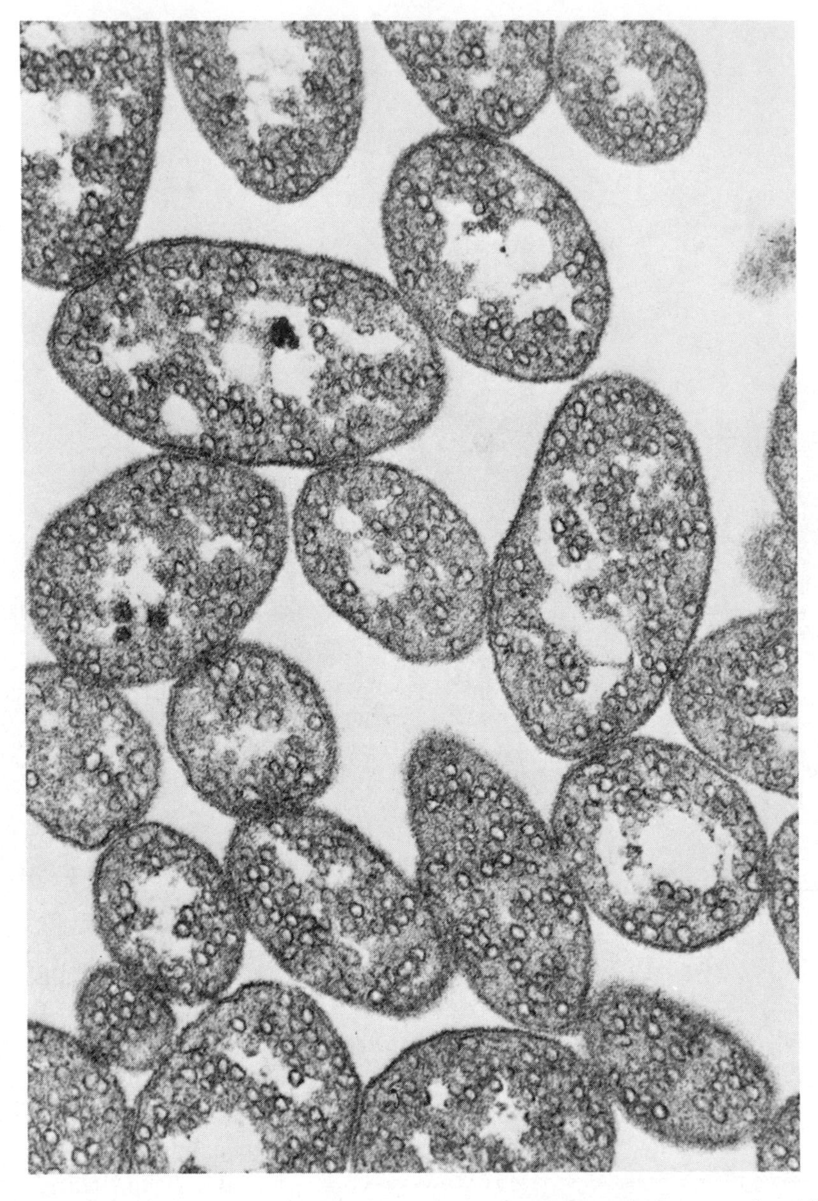

FIG. 8.8 Vesicle-like thylakoids ($\times$35,000) in *Rhodopseudomonas spheroides*. (W. Menke, 1966.)

regular in some areas where the membrane, apparently enveloping the lamella like a pillow-case envelopes a pillow, has been torn away. Park and co-workers dubbed the granular units "quantasomes," in the belief that they are the sites of the primary photochemical process—they called it "quantum conversion"—in photosynthesis. The size of the quantasomes is about $18 \times 16 \times 10.0$ nm; their upper surface is thus about 300 nm^2. Chlorophyll molecules in artificial monolayers, stacked obliquely, like books on a half-filled shelf, are known to occupy about 1 nm^2 each. (When put flat on their "faces," they cover about 2.4 nm^2 each.) Thus,

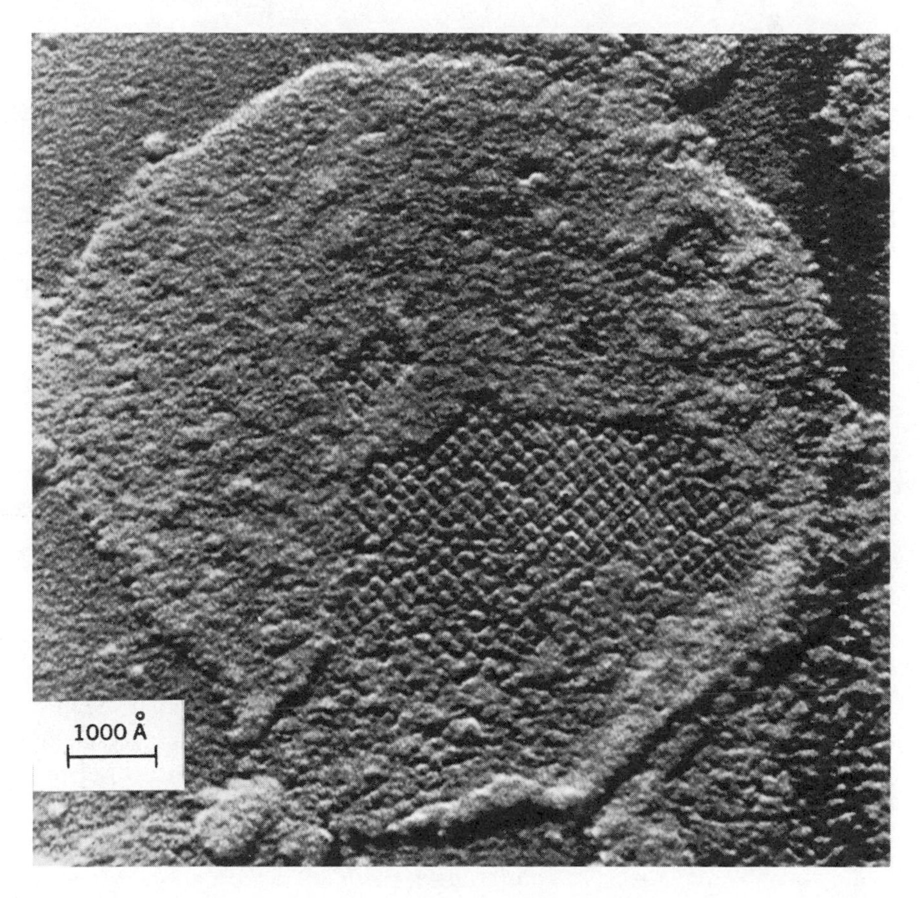

FIG. 8.9 "Quantasomes" on a chloroplast lamella. (R. B. Park and J. Biggins, 1964.) (1000Å = 1000 nm.)

a quantasome may offer just enough surface space to the about 300 chlorophyll molecules postulated to be present in a photosynthetic unit.

However, some recent experiments by E. N. Moudrianakis, (at Johns Hopkins University) suggested that "quantasomes" can be removed from the lamellae, like mushrooms cut off from a bed, and that the remaining lamellae can still reduce certain Hill oxidants in light. This argues against identification of "quantasomes" with complete photosynthetic units. However, it is not certain that Park and Moudrianakis had been looking at the same particles.

PROTEINS AND LIPOIDS

If one separates, by maceration and centrifugation, chloroplasts (or their fragments) from the rest of the cell material and analyzes them, one finds that, in contrast to the cytoplasm in which they were suspended (which is almost pure protein), chloroplasts contain 30–40 percent of nonproteidic material. This is classified as "lipoid" (fat-like), because it is insoluble in water but soluble in typical fat-solvents (such as alcohol and ether). Some of the chloroplast lipoids contain phosphate residues (phospholipids); others contain carbohydrates (glucolipids).

Lipoids are widespread in nature, and are known to perform two main biological functions. One is *energy storage,* associated with the high energy content of fats (that is, their high reduction level; see Chapter 5). The other is *selective permeability*. Lipoids are found in all membranes: the frog's skin, the membrane surrounding red blood corpuscles, the sheaths that enclose nerve fibers, etc. All such membranes contain 30–40 percent lipoid material, the rest being fibroid (threadlike, as contrasted to globular) proteins. A general picture has developed of membranes being built like plastic tablecloths, with a fibrous network underlying a plastic layer. The proteins provide the fibrous base, on which the amorphous lipoid layer is spread. This is the structure referred to above as the "unit membrane." Many such double layers can be stacked on top of each other, for example in nerve sheaths, chloroplasts, visual cones, and other cell organelles. All of them consist of alternating proteidic and lipoidic layers. In chloroplasts, the lamellae are closely packed inside the grana; outside the grana, they are separated by so-called "stroma" probably a pure proteidic medium (Fig. 8.4).

The function of membranes, when they are found surrounding a cell organ, a whole cell, or the organism as a whole, is to maintain inside a chemical composition different from that prevailing outside, while permitting a certain material exchange without which no life is possible. These membranes have so-called selective permeability, letting some molecules through, but preventing the passage of others. It is not enough to keep, let us say, large molecules inside (or outside) while letting smaller ones circulate freely, as could be done by a molecular sieve with holes of appropriate size. A distinction is made by many cell membranes between different simple molecules or ions, such as those of sodium and potassium. The membrane must thus contain a mechanism for selecting admissible from nonadmissible ions and molecules. More than that, it often has a mechanism for *active diffusion,* that is, for pumping certain molecules or ions in from the outside (and others, out from the inside) even *against a concentration gradient,* from lower to higher concentrations. (Natural or passive diffusion always goes in the opposite direction, as required by the entropy principle.)

The selective permeability of cell membranes is associated, at least in part, with the properties of thin lipoid layers. In some simple cases, the relevant factor is simply the solubility of various compounds in the lipoids. Those that are easily soluble penetrate the membrane easier than those that are practically insoluble. This is, however, not the whole story. To explain such phenomena as allowing the passage of potassium and barring the passage of sodium (or vice versa), special mechanisms have to be imagined. Since any motion against the gradient of concentration consumes free energy, it can occur only by coupling it with some free energy-supplying process, such as respiration (probably via high energy phosphate, ATP, as energy carrier). The details of the coupling mechanism remain to be elucidated, although some promising suggestions have been made in recent years, particularly by P. Mitchell in England.

The prime function of the protein layer in a membrane often may be to give it mechanical strength, because protein molecules consist of extended chainlike molecules, in which the links are strong chemical bonds and cannot be easily disrupted. These chains can form very stable mats, containing pores through which diffusion of even relatively large molecules can proceed. However, some proteins in the membrane must have a more active, catalytic function.

In chloroplasts, the stacked membranes may have another function. They offer support for a display of enzymes, permitting easy access of substrate molecules, directed passage of intermediates from one enzyme to another, and easy removal of the ultimate products. They provide, as it were, long metabolic conveyor belts. (The same picture applies to mitochondria, the site of several energy-releasing oxidation-reduction reaction steps in respiration.) A third function of membranes can be envisaged in chloroplasts (and, perhaps, also in visual rods): Absorbed light quanta must be efficiently utilized in them—in photosynthesis for photochemical energy storage and in vision for the initiation of electric signals in the optic nerve. Efficient propagation of *energy,* in addition to efficient transfer of *chemical entities,* may be needed to permit excitation to reach the site of its utilization.

The most important function of proteins in the living organism is to serve as catalysts in metabolic processes. Enzymes, the all-important biological catalysts, are such catalytic proteins; often they carry certain relatively small active ("prosthetic") groups directly involved in their catalytic action. All enzymes needed for the conversion of CO_2 to carbohydrates are located in the chloroplasts; some other enzymes (such as, *catalase* and *peroxidase*) have been found in them, too.

Certain colored proteins, such as the *iron*-porphyrin-containing *cytochromes,* have been found in chloroplasts and appear to serve as catalysts in photosynthesis. Like hemoglobin, they contain an iron atom in the center of a porphyrin molecule. (Other cytochromes play an important role in respiration in animals and plants.) More recently, a *copper*-containing, blue protein was found in chloroplasts and called *plastocyanin;* it, too, is supposed to have a catalytic function in photosynthesis. We shall return to these catalysts in Chapter 14. A *manganese*-containing enzyme has been implicated in photosynthesis, but no manganese-bearing protein has as yet been isolated from chloroplasts.

All chloroplast pigments seem to be attached to proteins. In the case of the (red and blue) phycobilins, the pigments can be easily and completely extracted, together with their carrier protein, into an aqueous medium.

Chlorphyll-protein complexes have been isolated from green bacteria, blue-green algae, and green plants; but not more than a small fraction of total chlorophyll was obtained in this form.

NUCLEOTIDES AND QUINONES

Another kind of nitrogen-containing basic life constituents, in addition to the long-known proteins, are the so-called nucleotides. In *nucleic acids*—the carriers of genetic information in the chromosomes—four types of nucleotides alternate in long chains, like letters in a script. Other nucleotides serve as catalysts in metabolic reactions. Chloroplasts contain two kinds of the latter—nicotinamide adenine dinucleotide (NAD^+), and nicotinamide adenine dinucleotide phosphate ($NADP^+$).

Among the "quinonoid" compounds present in chloroplasts, that is, compounds with a structure similar to that of simple quinones (see Chapter 5), and capable of reversible reduction to hydroquinone-like compounds, are the so-called *plastoquinones*.

We shall see in Chapters 14 and 17 that several above-mentioned compounds—cytochromes, plastoquinone, plastocyanin, and pyridine nucleotides—are assumed to act as intermediates in the path of hydrogen atoms (or electrons) from H_2O to CO_2 (see Figs. 5.4 and 14.4). They lose and acquire electrons reversibly, like members of a bucket brigade pass pails of water.

THE PIGMENTS: THEIR LOCATION IN THE CHLOROPLASTS

The chloroplast pigments are the most important specific components of the chloroplasts. They absorb light and initiate the chains of enzymatic reactions that lead to the conversion of H_2O and CO_2 into carbohydrates and oxygen. They are present in large amounts (up to 5% of the dry cell material by weight!). This is needed for photosynthesis to proceed in the practically available light fast enough to support the growth of the plant. (Because of the endergonic character of photosynthesis, no light-initiated chain reactions are possible!)

Because of this importance, the pigments will be discussed in a separate chapter (Chapter 9). Here, we shall only deal with their location in the chloroplasts. Unfortunately, electron micrographs tell us nothing about the location of pigments. Electrons do not distinguish between colored and colorless material, but only between materials of different density of the electronic cloud, that is, between lighter and heavier atoms. (Higher atomic number means a higher charge of the nucleus, and thus also a denser crowd of electrons around it!) The sharp relief in Fig. 8.3 was obtained by "staining" the preparation, by depositing on it heavy metal atoms from an atomic beam (created by heating a metal grain in vacuum). In other electron-micrographs, the contrast is achieved by means of heavy osmium atoms precipitated by chemical reduction of osmium tetroxide (osmic acid, OsO_4). The osmium atoms are precipitated preferentially, in (or on) layers of a stronger reducing power. It is not quite certain yet what these layers are, but it does seem likely that the darker layers (see Fig. 8.4) are more lipoidic in nature, and the lighter ones, more proteidic.

In any case, even the best electron micrographs tell us nothing about the location, in the layered structure, of the molecules of chlorophyll or of other plant pigments.

A plausible picture of this location can be based on the consideration of the structure of the chlorophyll molecules, and its likely behavior in a system of alternating proteidic and lipoidic layers. These layers are hydrophilic and hydrophobic, respectively. Water molecules are electric dipoles, with a positive charge on the hydrogen side and a negative charge on the oxygen side of the molecules. They are, therefore, attracted by free ions, which, in contact with water, surround themselves by a halo of water molecules (Fig. 8.10*a*), and also by ionized or polarized spots in large molecules. In the latter case, water dipoles crowd around the polar spot like wasps around a drop of honey (Fig. 8.10*b*). Protein molecules contain many charged spots, both negative (acidic) and positive (basic). Therefore, they are more or less strongly hydrophilic (water-attracting). Many protein molecules are, in fact, soluble in water; that is, they attract so many water molecules that the latter bodily lift and carry them into the aqueous medium. A proteidic layer is, therefore, hydrophilic. Lipids (fats), on the other hand, are hydrophobic (water-repellent), because they contain no polar groups and exercise only general "van der Waals" attraction forces on all neighboring molecules. We cannot discuss here the nature of the van der Waals forces. (F. London has interpreted them as being due to the fact that even uncharged molecules consist of moving positive and negative particles;

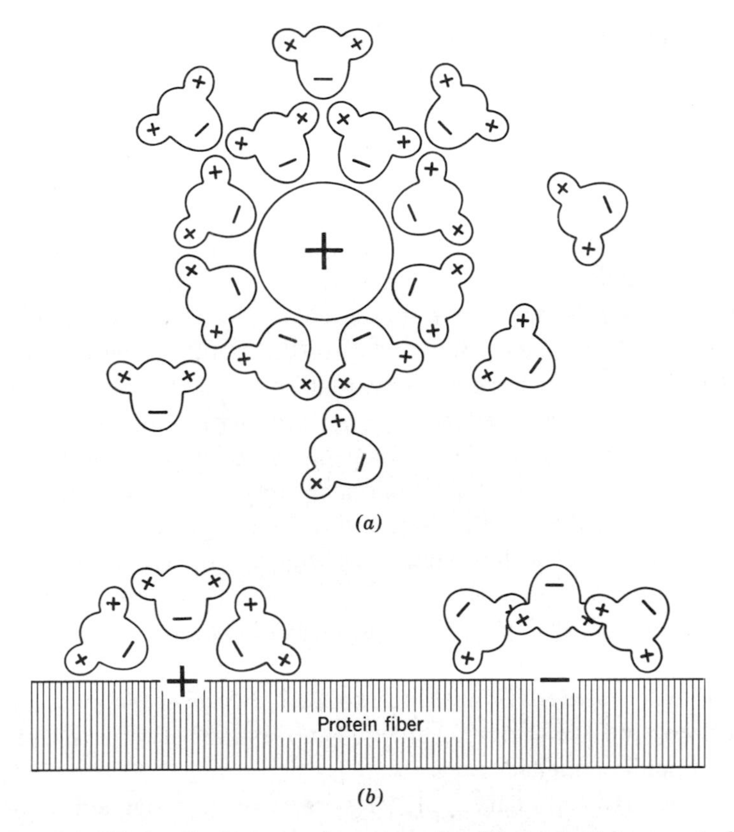

FIG. 8.10 (*a*) Water dipoles around an ion. (*b*) Water dipoles around charged spots on a protein molecule.

when two such molecules approach each other, these particles are displaced and the originally nonpolar molecules become "dipoles"—and dipoles attract each other.) In competition between polar water molecules and nonpolar organic molecules for positions close to other nonpolar organic molecules, the second ones win out, the water molecules are squeezed out. It is not that such "hydrophobic" molecules "dislike" water, but they like organic molecules more!

Phospholipids are lipoids that contain phosphoric acid anions; these *are* polar, so that water molecules are attracted to them. In other words, hydrophobic molecules may contain hydrophilic spots. The overall behavior of such molecules is the balance of their hydrophobic and hydrophilic properties. Thus, *soaps* and *detergents* attach themselves to fat molecules by their hydrophobic parts, and carry them into water by means of their hydrophilic parts.

When a molecule that contains both hydrophilic and hydrophobic parts is brought into a structure consisting of alternating hydrophilic and hydrophobic layers, one expects it to find a most stable position at the protein-lipoid *interface,* where it can satisfy both its attraction to water and its affinity for organic molecules.

We know (see Chapter 9) that the chlorophyll molecule is built like a tadpole, with a roughly square head (the porphin ring) and a long phytol tail. The head is polar because it contains the magnesium atom, which tends to acquire a positive charge, making the rest of the molecule negative, and this polarity causes an affinity to water. The phytol tail, on the other hand, is nonpolar and therefore hydrophobic. Chlorophyll molecules should therefore accumulate on lipoid-protein interfaces, with their flexible phytol tails dipping into the lipoid layers, and the rigid porphin head attracted to the aqueous layer. Chlorophyll may thus form a one-molecule thick layer on the interface between the lipoid and the protein lamellae in chloroplasts (Fig. 8.11). Optical studies of such properties of chloroplasts as birefringence and dichroism support the idea that chlorophyll molecules form separate layers between the proteidic and the lipoidic lamellae.

Preparing artificial pigment monolayers by evaporating a drop of chlorophyll solution in petroleum ether spread on the surface of water, can lead to two types of layers, crystalline and amorphous. Crystalline chlo-

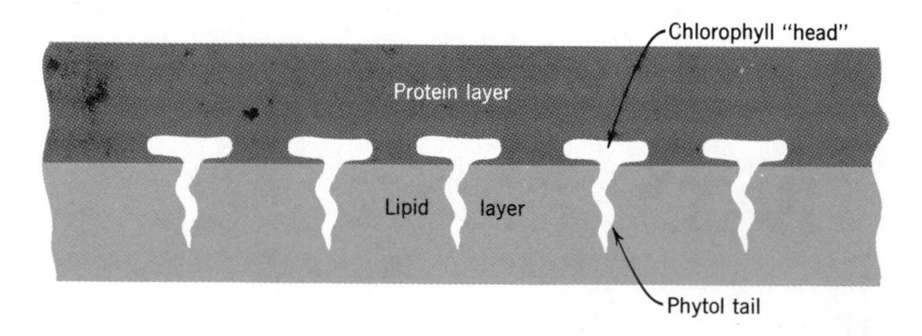

FIG. 8.11 Chlorophyll molecules on the interface between lipid and protein lamellae.

rophyll layers are characterized by greater surface density, and by a considerable shift in the position of the red absorption band—from 660–670 nm in solution, to 730–740 nm, probably caused by molecular interactions in a closely packed regular two-dimensional lattice (see absorption bands a and d in Fig. 8.12). Noncrystalline chlorophyll monolayers are less dense, and have absorption bands only slightly shifted from their position in solution (to 670–680 nm; see Fig. 8.12*b*). The absorption spectra of pigments in live cells show no evidence of crystalline structure; the red absorption band of chlorophyll *a* in the living cells is located at 670–680 nm (Fig. 8.12*c*). Thus, if chlorophyll monolayers *are* present in chloroplasts, they must be of amorphous, rather than crystalline kind. Some evidence has been found in recent years that a small fraction of chlorophyll *a* in vivo may have a more orderly arrangement than the bulk of it. These may be the molecules

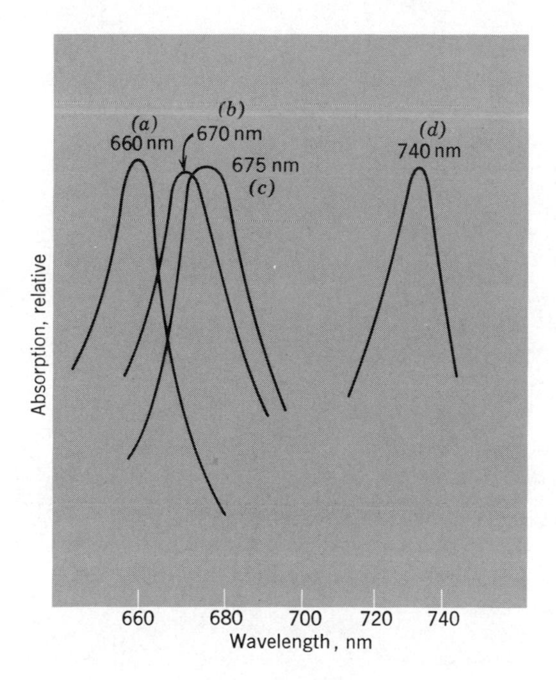

FIG. 8.12 "Tops" of the red absorption band of chlorophyll *a*, in true solution (*a*); in colloidal solution or amorphous monolayer (*b*); in living cell (*c*); and in a crystalline monolayer (*d*). Note greater width of the band in (*c*).

most closely associated with the "reaction centers" in the photosynthetic units.

The alternation of lipoidic and proteidic layers on two sides of chlorophyll layers may permit the two primary reaction products, the strong oxidant and the strong reductant, to escape from each other, and thus prevent a back reaction between them.

We mentioned before the existence of "photosynthetic units" often containing about 300 Chl *a* molecules in green plants and about 50 bacteriochlorophyll molecules in bacteria. How can the existence of such units be reconciled with the concept of extended monomolecular pigment

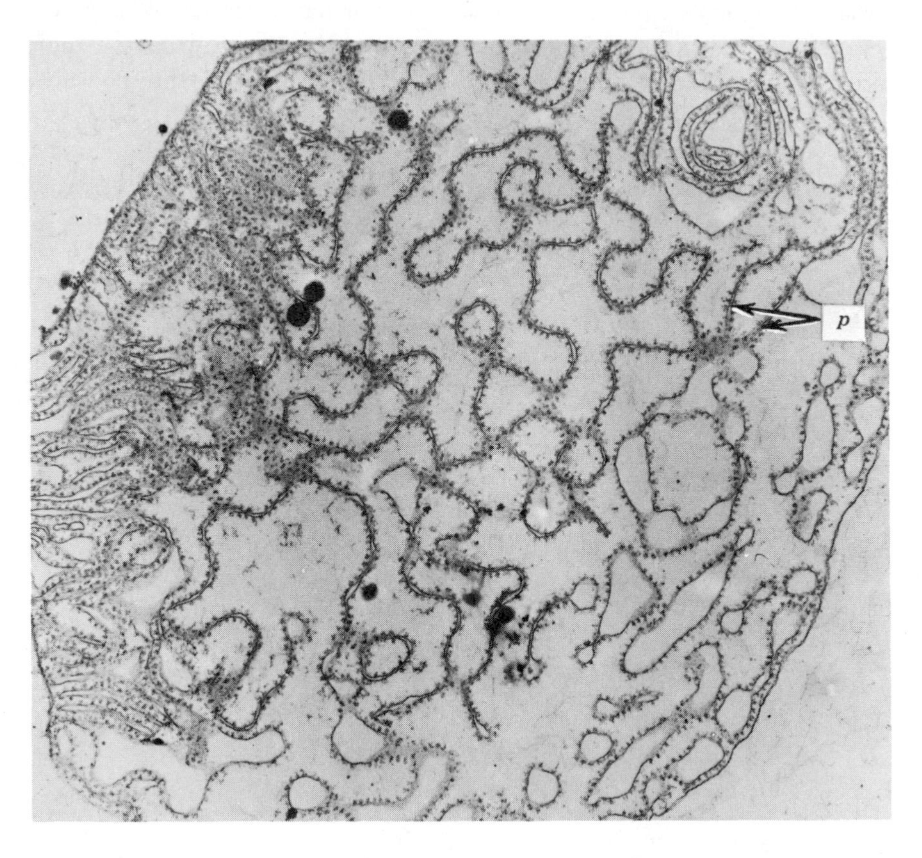

FIG. 8.13 *Phycobilisomes* (*p*), in chloroplast of *Porphyridium cruentum*. (E. Gantt and S. F. Conti, 1966.)

layers? Perhaps, the units are "cobblestones" in the "pavement" of the lamellae. This point remains to be elucidated.

Even if we assume that chlorophyll molecules are arranged in monomolecular layers on protein-lipoid interfaces, many questions remain unanswered, such as: Are chlorophyll *a* and chlorophyll *b* mixed indiscriminately in this green coat? Where are the carotenoids? Where are the phycobilins? Recent electron micrographs by E. Gantt and S. F. Conti at the Dartmouth Medical School have shown small phycobilin-containing granules, approximately 35 nm in diameter, attached to the chloroplast lamellae in the red alga *Porphyridium* (Fig. 8.13). The name *phycobilisomes* has been suggested for them. Here again, the problem arises of relation between such three-dimensional structures and the postulated monomolecular pigment layers.

Detailed models of the arrangement of the pigments and other components in the chloroplast lamellae have been constructed by many biologists, but they remain speculative.

Finally, the concept (see Chapters 13–17) of two separate photochemical systems (and thus, presumably, also of two types of photosynthetic units, of equal or different size) being involved in photosynthesis, remains to be reconciled with the pictures presented in this chapter. We shall deal in Chapter 16 with experiments suggesting that units of the two types can be separated, at least partially, by certain fractionation procedures.

Chapter 9

The Photosynthetic Pigments

THE PIGMENT MOLECULES

The most important components of the chloroplast lamellae are colored organic compounds—the photosynthetic pigments. Already Ingenhousz noted, in 1795, that colorless plant organs cannot "improve the air."

A rule, almost too obvious to deserve the title of the "first law of photochemistry" (which is sometimes bestowed on it), is that only light absorbed in a reacting system can have a chemical effect. However, when light absorption initiates a long chain of exergonic reactions, small, "catalytic" amounts of pigments may suffice. (An example is the photoinduction of flowering and other "photomorphological" effects, where the primary light-absorbing component proved very elusive, and was only recently identified as a colored protein, the "phytochrome.") Much larger amounts of pigments must be present to produce a sizable photochemical reaction when such chains cannot develop. A chain reaction is out of the question in photosynthesis because of the endergonic nature of the overall reaction, and yet photosynthesis must be a very fast process to provide all the organic material needed for the growth and life activity of the plant. This is why photosynthesizing cells have to contain very large amounts of pigments (up to 5% or more of total dry material).

Pigments are molecules strongly absorbing visible light. Such absorption is restricted to certain classes of compounds. In painting, we use mostly the cheaper and more stable inorganic pigments (ochre, cobalt blue, cinnabar, chrome yellow, etc.); but good organic pigments have one or two orders of magnitude higher molecular absorption coefficients than the best inorganic ones.

The molecular structure of such organic pigments is characterized by long chains (or closed rings) of so-called "conjugated" double bonds. Carbon atoms, being quadrivalent, can be bound, in organic molecules, by single, double, or triple bonds, of which only single and double bonds are encountered in complex molecular structures. (The four bonds of a carbon atom are directed from the center to the four corners of a tetrahedron, and deflection from these directions needed to form a triple bond causes a stress.) Rings (or linear chains) consisting of regularly alternating ("conjugated") single and double bonds are particularly stable because they are reinforced by "resonance" (for example, that between the structures —C=C—C=C— and =C—C=C—C=)—interaction between several structures with the same arrangement of atoms, but different distribution of electrons. The most common of conjugated *rings* is that present in benzene, C_6H_6. It contains three single and three double bonds, and two main resonating structures are possible  and  (omitting the C and H symbols). More complicated rings of the same character are present in other "aromatic" compounds, for example, napthalene, $C_{10}H_8$, 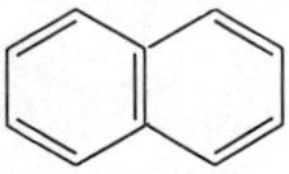 and anthracene, $C_{14}H_{10}$, 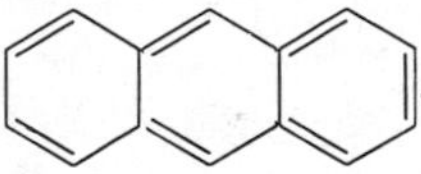 . (Only one of the resonating structures is represented.)

The nitrogen atom, being trivalent, can form a single bond with one carbon atom, and a double bond with another; it can be, therefore, substituted for a carbon atom (or more precisely, for a CH-group) in conjugated systems. Rings containing atoms other than C are called *heterocyclic* (in contrast to the *homocyclic* rings, containing only carbon atoms). The simplest of them is pyridine, C_5H_5N, or N. One heterocyclic ring, common in plant pigments, is the five-membered pyrrol ring: N.

It is typical of molecules containing chains or rings of conjugated bonds to have strong absorption bands at relatively long waves, that is, in the visible or in the near ultraviolet. (Organic molecules containing only single bonds, or isolated double bonds, usually absorb in the far ultraviolet.) As the number of conjugated double bonds increases, absorption shifts towards the longer waves, and becomes stronger. Thus, in benzene, the first absorption band lies below 250 nm, while in anthracene, it lies close to the visible, making this compound yellowish. In still more complex conjugated ring systems, such as those present in *indigo* or in *chlorophyll* (or in long linear conjugated chains, such as those found in *carotenoids*), the absorption bands shift into the visible, making these compounds pigments.

The strong light absorption by conjugated systems has to do with their resonance structures (of which the two above-mentioned energetically equivalent structures of benzene are the prototype). As shown in Fig. 9.1, the splitting of excited electronic energy levels into several

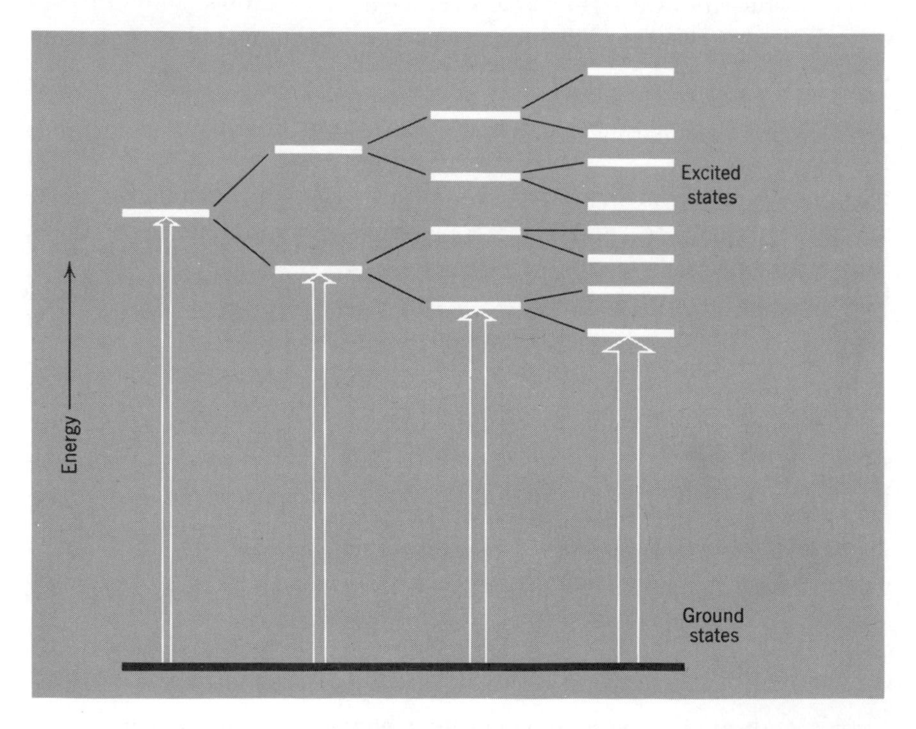

FIG. 9.1 Splitting of excited energy level by resonance in a chain of identical atoms (or molecules).

components, due to the existence of these resonating structures, leads to narrowing of the gap between the ground state and the first excited state; the first absorption band thus moves towards longer waves.

CHLOROPLAST PIGMENTS: GENERAL REMARKS

Photosynthetic organs of plants always contain an assortment of pigments. These can be divided into three major classes (see Tables 9.1–9.3): (1) chlorophylls and (2) carotenoids—both water insoluble, and (3) phycobilins—water-soluble (because of their attachment to water-soluble proteins).

The *first* general feature of the distribution of chloroplast pigments is the universal presence of *chlorophyll a,* found in *all* photosynthesizing cells (except photosynthetic bacteria, where a related pigment bacteriochlorophyll (BChl) is found instead). Chlorophyll *a* is present also in brown, red, and blue-green algae, where its green color is masked by other pigments. So far, no plant was found capable of true photosynthesis which did not contain chlorophyll *a* (Chl *a*). The widespread occurrence, as well as certain chemical properties of chlorophyll *a* in vivo (and of bacteriochlorophyll in bacteria), suggest that these pigments (or, at least some of their molecules) play an active role in photosynthesis, functioning as *photoenzymes.* All other pigments (and a large part of Chl *a* or BChl itself) seem to serve only as physical energy suppliers. Pigments other than Chl *a* are often called *accessory* pigments.

An enzyme is a protein catalyst that mediates a chemical reaction by active but reversible participation in it—as a middleman who buys a commodity from the seller to sell it to the buyer. The commodity, in the case of a redox reaction, is hydrogen atoms or electrons. A photoenzyme is a broker who can ply its trade only after having been activated by light absorption. Because of its role as a light-activated enzyme in photosynthesis, chlorophyll has been called "the most important organic compound on earth."

The *second* common, but not universal rule is the presence of a second kind of chlorophyll, in addition to Chl *a*. Higher plants and green algae contain, in addition to the yellow-green chlorophyll *a*, the blue-green

TABLE 9.1 The Chlorophylls

Type of chlorophyll	Characteristic absorption peaks		Occurrence
	In organic solvents, nm	In cells, nm	
Chl *a*	420, 660	435, 670–680 (several forms)	All photosynthesizing plants (except bacteria)
Chl *b*	453, 643	480, 650	Higher plants and green algae
Chl *c*	445, 625	Red band at 645	Diatoms and brown algae
Chl *d*	450, 690	Red band at 740	Reported in some red algae
Chlorobium chlorophyll (also called "bacterio-viridin")	Two forms: (1) 425, 650 (2) 432, 660	Red band at 750 (or 760)	Green bacteria
Bacteriochlorophyll *a* (BChl*a*)	365, 605, 770	Red bands at 800, 850, and 890	Purple and green bacteria
Bacteriochlorophyll *b* (BChl*b*)	368, 582, 795	Red band at 1,017	Found in a strain of *Rhodopseudomonas*, (a purple bacterium)

chlorophyll b (Chl *b*)—a locally oxidized derivative of Chl *a*. Several other chlorophyll-type compounds have been described in the literature. "*Chlorophyll c*" is found in brown algae and diatoms. A "chlorophyll *d*," identified by H. Strain and others in red algae, has proved elusive in later experiments. A "chlorophyll *e*" has been reported in golden-yellow algae.

Green bacteria contain mostly so-called "Chlorobium chlorophyll," or *bacterioviridin,* in addition to traces of *bacteriochlorophyll,* which is the main pigment of purple bacteria.

TABLE 9.2 The Carotenoids

Types of carotenoids	Characteristic absorption peaks, nm[a]	Occurrence
	I. CAROTENES	
α-Carotene	In hexane, at 420, 440, 470	Many leaves, and certain algae. In red algae and a group of green algae called *Siphonales*, it is the major carotene
β-Carotene	In hexane, at 425, 450, 480	Main carotene of all other plants
γ-Carotene	In hexane, at 440, 460, 495	Major carotene of green sulfur bacteria; traces in some plants
	II. CAROTENOLS (also called "xanthophylls")	
Luteol	In ethanol, at 425, 445, 475	Major carotenol of green leaves, green algae and red algae
Violaxanthol	In ethanol, at 425, 450, 475	Second major carotenol of leaves
Fucoxanthol	In hexane, at 425, 450, 475	Major carotenol of diatoms and brown algae
Spirilloxanthol	In hexane, at 464, 490, 524	Common in purple bacteria

[a] It has been difficult to establish the exact location of carotenoid bands in vivo (except in the case of purple bacteria) because of their strong overlapping with the blue-violet bands of chlorophylls. The bands in vivo are estimated to be shifted by about 20 nm to the long-wave side from their position in solution. In the case of fucoxanthol, absorption extends in vivo to 550 nm.

TABLE 9.3 The Phycobilins

Types of phycobilins	Absorption peaks	Occurrence
Phycoerythrins	In water, and *in vivo:* 490, 546, and 576 nm	Main phycobilin in red algae, also found in some blue-green algae
Phycocyanins	At 618 nm, in water and *in vivo*	Main phycobilin of blue-green algae; also found in red algae
Allophycocyanin	At 654 nm in phosphate buffer at pH 6.5	Found in blue-green and red algae

An additional complication—to be discussed later in the chapter—arises from the apparent presence in living cells of several forms of one and the same pigment, which become identical upon extraction. This fact was long known in the case of bacteria, where three distinctly separated far-red absorption bands of bacteriochlorophyll are observed in vivo, at approximately 800, 850 and 890 nm, but give rise, upon extraction into an organic solvent, to a single absorption band at 770 nm. It has been surmised that bacteriochlorophyll is present in the living cell in the form of complexes with three different proteins, or in three different aggregation stages. A similar but less conspicuous polymorphism exists also in the case of chlorophyll *a* in green plants.

The *third* general rule, from which no exceptions are known, is the presence in all photosynthesizing cells of an assortment of carotenoids—relatives of the pigment that accounts for the orange color of carrot roots. Most carotenoids are yellow or orange. Their color is normally masked by chlorophyll, but in chlorophyll-deficient, so-called *aurea* varieties of plants, or in fall, when chlorophyll disintegrates, the yellow pigments become visible. Some of them become in autumn orange, or even red, through oxidation, thus contributing to the variety of foliage colors. (However, important contributions to the bright red or purple colors of some leaves in autumn are also made by pigments of another class, so-called anthocyanins.)

The specific assortment of carotenoids is different in plants of different classes. In general, there are two major groups: the *carotenes* (which are hydrocarbons) and the *carotenols*. The ending "ol" suggests that these compounds are mostly alcohols (although some of them are ketones). The plant carotenols are also designated as *xanthophylls* (from *xanthos* = orange-yellow).

Certain algae are brown; examples are large algae such as *Fucus*, familiar to all visitors on ocean beaches, and the free-swimming microscopic diatoms (*Bacillariophyceae*), having rigid silica skeletons. The diatoms are among the most successful plants on earth; they fill the upper layers of the oceans and may account for as much photosynthesis on earth as the green land plants. The brown color of all these organisms is due to a special carotenol, called *fucoxanthol*.

The *fourth* important fact is the presence, in the red marine algae (*Rhodophyceae*), and the primitive blue-green algae (*Cyanophyceae*, encountered on land as well as in shallow water), of pigments of a still different class, called *phycobilins* (from *phycos* = alga, and *bilin* = bile pigment), because of their similarity to the pigments of the bile. These pigments may be present in amounts equal to, or greater than, those of chlorophyll *a*. Phycobilin is a generic name for pigments of two kinds: the red *phycoerythrins* (from *erythros* = brick red), and the blue phycocyanins (from *cyanos* = blue). The phycobilins are structurally related to chlorophylls (see below). The phycoerythrins appear to have the definite function of improving the light absorption in the middle of the visible spectrum; the same is probably true of the above-mentioned "brown" carotenoid, fucoxanthol.

Phycobilins can be extracted from the algae simply by placing the latter in distilled water, while chlorophyll extraction requires organic solvents. We have seen that chloroplasts contain hydrophilic proteins and hydrophobic lipoids; the latter, like grease spots, can be extracted only by organic solvents. As mentioned in Chapter 8, chlorophyll molecules have affinity to both polar and nonpolar molecules; this is why they can be most easily extracted by a mixture of organic solvents, such as acetone or alcohol, with water. The latter disintegrates the protein structure, while the former pulls out the lipoid molecules.

Phycobilins have no phytol chains and are attached to water-soluble proteins; they are thus easily extractable into pure water. Most carotenoids have no affinity for water and are soluble only in organic solvents.

CHLOROPHYLLS

Structure

The chlorophylls *a* and *b* have a common basic structure. Some familiarity with organic chemistry is needed to discuss it. It is a "porphyrin" structure, consisting of four pyrrol rings (Fig. 9.2), joined into a single master ring by CH "bridges." Figure 9.2 represents *porphin,* the mother substance of all porphyrins. Often a metal atom is found in the center of a porphyrin—iron in heme, magnesium in chlorophyll. In the latter, a long chain of carbon atoms ("phytol chain," from *phytos*=plant) is attached to the ring system. Other groups attached to the porphin skeleton will be discussed below. These side chains, and

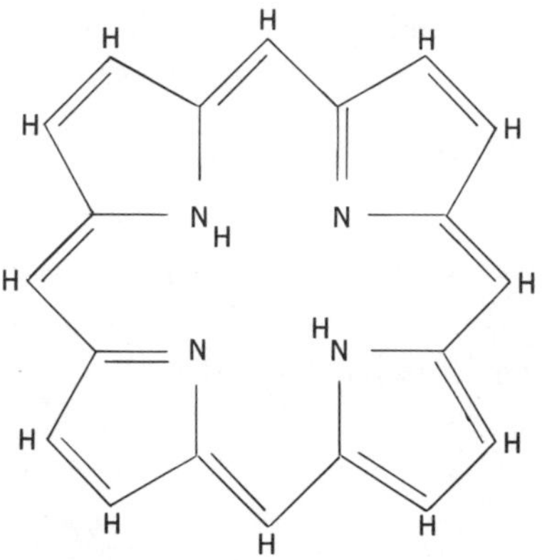

FIG. 9.2 Porphin structure consisting of four pyrrol rings (all corners except those marked N are occupied by carbon atoms).

the double bonds present in the overall structure, distinguish among themselves the numerous compounds of this class.

The porphin unit plays a very important role in nature. It forms the skeleton of both chlorophyll and heme, the red pigment of blood, and of other compounds of physiological significance (for example, the cytochromes).

The complete molecular structure of *chlorophyll a* is represented in Fig. 9.3. (Two Nobel prizes were awarded for the development of this structure, to the German chemists Richard Willstätter, and Hans Fisher, respectively; a third one was awarded to Robert B. Woodward of Harvard for total in vitro synthesis of chlorophyll.) Compared to the porphin structure of Fig. 9.2, that of chlorophyll *a* is characterized by a missing double bond in one of the pyrrol rings (ring IV). It is thus derived from a dihydroporphin (a porphin containing two additional hydrogen atoms).

Bacteriochlorophyll is derived from tetrahydroporphin, with *two* fewer double bonds and *four* more hydrogen atoms than porphin.

A peculiar and apparently important characteristic of chlorophyll (and of the bacteriolchlorophyll, too) is the presence of a fifth, "homocyclic," five-membered ring (ring V in Fig. 9.3) which carries a carbonyl group, C=O. This additional ring may be the "nerve center" of the chlorophyll

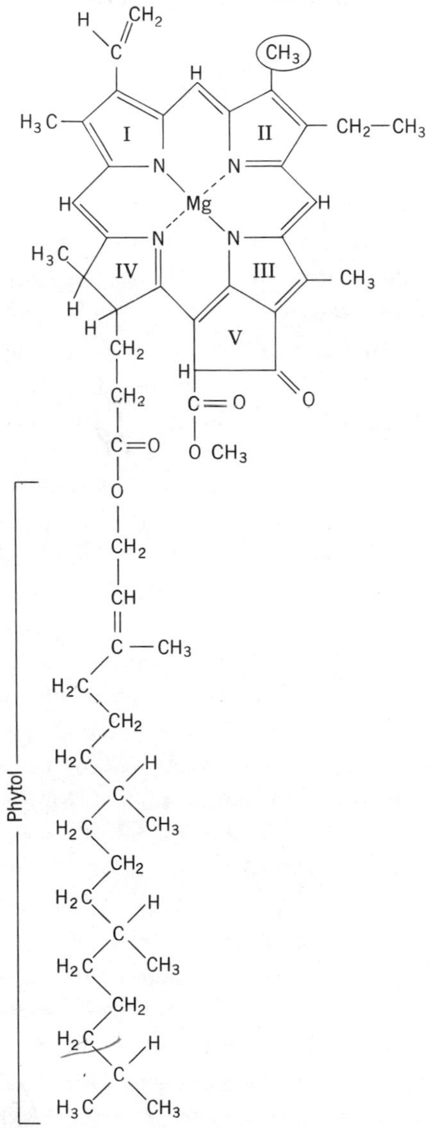

FIG. 9.3 Chlorophyll *a* (the circled CH_3 group is replaced by CHO group in chlorophyll *b*). All corners except those marked N are occupied by carbon atoms.

molecule, most closely associated with its photocatalytic action in photosynthesis. This conclusion is not based on direct evidence, but rather on general feeling arising from the totality of chemical and photochemical experience with chlorophyll and related compounds.

Another peculiarity of the chlorophyll molecule is the above-mentioned presence of a magnesium atom in the center of the molecule. We do not know its function. In hemin, the same position is occupied by an atom of iron, and the capacity of iron to exist in two oxidation states, as ferric ion, Fe^{+3} and ferrous ion, Fe^{+2}, gives a hint as to its possible catalytic function. Magnesium does not have a similar property, and its special appropriateness to serve as center of such a catalytically active molecule is not clear. (Of all commonly available metals, it may perhaps fit best into the available space!) The presence of a metal atom, with its tendency to assume a positive charge, pushes the electrons in the porphyrin ring towards the periphery of the molecule, and thus into ring V. This may be important for its catalytic properties. But, again, this is not a matter of clear-cut evidence, but of chemical "feeling."

Some short side chains in the chlorophyll molecule may be simply residues left over from its synthesis in the organism, which begins with short-chain molecules, such as those of acetic acid, CH_3COOH, and of glycine, NH_2CH_2COOH. When these are combined into rings, methyl groups, CH_3, are left hanging outside the ring. However, the presence in ring I of a nonsaturated (vinyl) side chain, $—CH{=}CH_2$, may be of some significance because its double bond is conjugated on one side with the ring system, and thus affects its overall properties. Also important is the presence, in ring IV, of the long and almost saturated phytol chain. (Phytol contains only one double bond, and fourteen single bonds.) This chain causes chlorophyll molecules to attach themselves to other saturated long-chain molecules in the chloroplasts, that is, it gives it *lipoid solubility* (lipoids being a generic name for "fatlike" compounds). The general rule in organic chemistry is that polar molecules associate with other polar molecules or groups (and are, therefore, attracted to water, which is the most polar of all simple solvents), while nonpolar chains associate selectively with other nonpolar chains. (In inorganic chemistry, on the other hand, the most important bonds involve *positive* ions associating themselves with *negative* ions, as in the formation of acids, alkalis, and salts.)

Chlorophyll b the second variety of chlorophyll present in green plants and green algae—differs from chlorophyll a only by local oxidation of one side chain in ring II (marked by a ring in Fig. 9.3). This side chain is $-CH_3$ in Chl a and $-\underset{\underset{O}{\|}}{C}-H$ in Chl b.

Absorption Spectra

As a pigment, chlorophyll a is characterized by two strong absorption bands, located in the blue-violet and in the red region of the spectrum (Fig. 9.4a). The first of them, often called the *Soret band,* is common

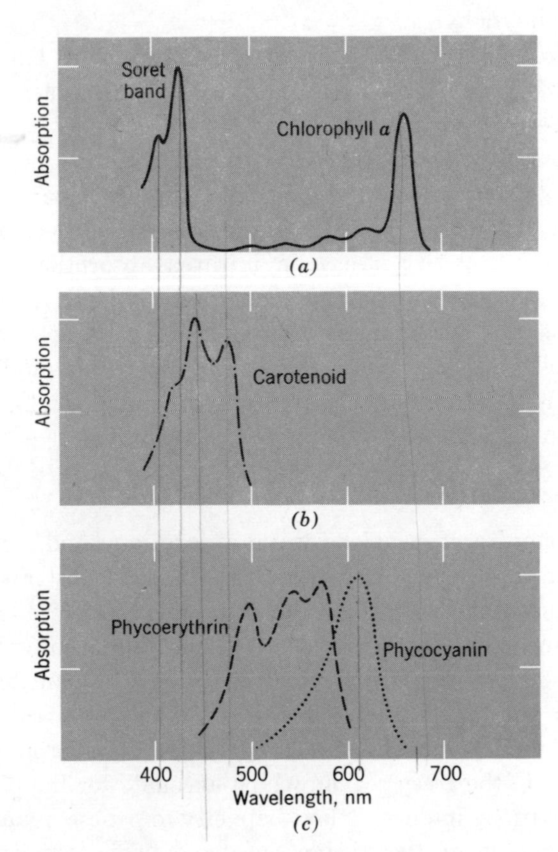

FIG. 9.4 Absorption spectra of three types of chloroplast pigments. (a) chlorophylls; (b) carotenoids; (c) phycoerythrins and phycocyanins.

to all porphin derivatives; but the second one is peculiar of chlorophyll and other compounds derived from dihydroporphine. In diethyl ether, the first has its maximum at 430 nm and the second at 660 nm. In the living cell, these bands lie at about 440 and 675 nm, respectively, and are complex (see last section of this chapter). The absorption is weak between the two bands, that is, in the green part of the spectrum, causing the green color of vegetation.

Chlorophyll *b* spectrum differs from that of Chl *a* by the two bands being closer together: the Soret band of Chl *b* is located (in ether) at 453 nm, and the red band at 643 nm. The ratio of intensities of the two bands is shifted in Chl *b* strongly in favor of the Soret band compared to chlorophyll *a*.

Bacteriochlorophyll *a* (BChl *a*) also has two main bands, but they lie farther apart; the Soret band in the near ultraviolet (at 365 nm) and the long-wave band, at the limit of the visible, at 770 nm (in methanol). In the living cell, the same bands lie at about 367 and 800–900 nm. The second one is usually split into three components, at about 800, 850, and 890 nm. A bacteriochlorophyll *b* has been recently discovered in some bacteria; it has an infrared absorption band at 1014 nm. Another chlorophyll derivative (mentioned before) is found in green bacteria; it is the "Chlorobium chlorophyll" or "bacterioviridin." In the living cell, the long-wave bands of this chlorophyll are found at 750 or 760 nm. Chlorobium chlorophyll-carrying green bacteria always contain also small quantities of bacteriochlorophyll *a*.

Chromatic Adaptation

The presence of a strong, long-wave absorption band, covering much of the red and orange region, in which sunlight is most abundant, may be one of the reasons why chlorophyll has been selected by nature as the main pigment in photosynthesis. Without this absorption band, the parts of the solar spectrum that contain the greatest amounts of useful energy would not be efficiently absorbed by cells. On the other hand, the specific form of the absorption spectrum of chlorophyll, with its absorption gap in the green region, which accounts for the pleasant green color of vegetation, implies a low capacity of these plants to utilize light in the middle of the visible spectrum. This deficiency is made up, at least to some extent, by the development of tiered vegetation: one tier of leaves lets enough light through for a second and a third layer to survive. In deciduous forests, bushes and grass can live on the ground. (Coniferous forests, on the other hand, often have such dense crowns of needles that the shadow under them is too deep for any vegetation to develop on the ground.)

The position of the absorption bands in the red cannot be the only reason why chlorophyll *a* had been selected by evolution as *the* photocatalyst in photosynthesis. Other pigments exist that would provide a better coverage of the visible spectrum. Additional reasons for this selection must be sought in specific *physicochemical* properties that bind chlorophyll in appropriate locations in the cell; and even more, in specific *catalytic* properties, which permit it to serve as an effective photocatalyst in photosynthesis. The selection of chlorophyll may be the best solution nature found to satisfy two needs: (1) for effective absorption of visible light, and (2) for proper photocatalytic properties.

As mentioned above, green land plants and green algae contain chlorophyll *b* in addition to chlorophyll *a*. Because of the shift of its Soret band, chlorophyll *b* in solution is blue-green, while chlorophyll *a* is yellowish green. The presence of chlorophyll *b*, therefore, narrows somewhat the "green gap" in the absorption spectrum of leaves; this may be the rationale for its presence, in relatively larger quantities, in shade-adapted plants.

We have also encountered another method of plant adaptation to life under water. The brown algae, as well as the diatoms, narrow the green gap by the provision of two other pigments—chlorophyll *c*, (whose chemical nature is not yet well known; it may be a porphin rather than a chlorin derivative), and of a specific carotenoid pigment, fucoxanthol (to be discussed below). Even more effective in filling the "green gap" is phycoerythrin, the main pigment of red algae (cf. below).

PHYCOBILINS (PHYCOERYTHRINS AND PHYCOCYANINS)

Structure

The chemical structure of the phycobilins is related to that of chlorophylls. If one snips the ring system of chlorophyll and permits the four pyrrol rings to straighten out, the magnesium atom slips out and what remains is an open conjugated system of four pyrrol rings. This is the fundamental structure of the *bilin* pigments, which can be considered as derived from a common parent, the hydrocarbon *bilan* (see Fig. 9.5*a* and *b*). The bilins are so called because they were first discovered in bile, to which they give its dark color. They may be, in fact, products of metabolic transformation of hemoglobin and chlorophyll, ingested with animal and plant food. The name phycobilin means algal bilin. Recently, it has been shown that the pigment phytochrome, responsible for the light control of seed germination and of flowering, is chemically related to the phycobilins.

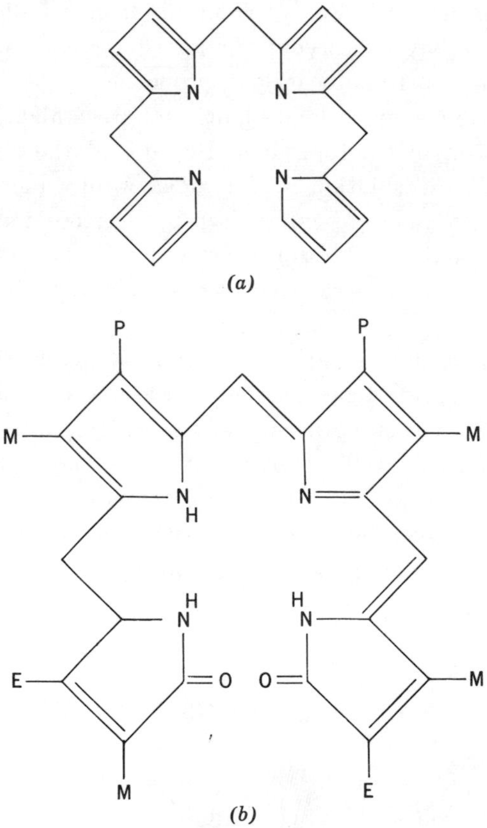

(a)

(b)

FIG. 9.5 Structure of bilin pigments. (*a*) bilan; (*b*) phycoerythrobilin. (All unmarked corners are occupied by carbon atoms.) (E, ethyl; M, methyl; P, propionyl groups.)

Absorption Spectra

In algae living deep under the sea, the deficiencies of chlorophyll as a light absorber become critical because light reaching these organisms is filtered through thick greenish-blue layers of water. (Dissolved salts, and organisms living in the surface layers, contribute to the absorption of red and blue-violet light.) Plants living under those conditions have evolved auxiliary pigments that absorb green light. Red algae, in particular, have solved the difficulty by means of a phycobilin called phycoerythrin, with absorption bands in the middle of the visible spectrum. For example, one type of phycoerythrin has bands at about 500, 545 and 570 nm (Fig. 9.4*c*). This permits these algae to perform photosynthesis (and thus to grow) in dim blue-green light prevailing in the depths of the ocean. Together, chlorophyll and phycoerythrin let through only a band in the far-red. This is why the algae which contain phycoerythrin appear red.

The deeper under the sea a red alga lives, the more phycoerythrin it contains in relation to chlorophyll.

However, complications make the overall picture more complex. For example, some algae encountered in great depth are green. They manage to survive by slowing down their life processes sufficiently to permit the small amount of light that their green pigment is able to catch to cover their needs.

Some surface, or even terrestrial, algae contain a pigment related to phycoerythrin, called *phycocyanin*. This is blue rather than red because its main absorption band is located at about 630 nm (Fig. 9.4*c*). It does not fill up the gap in the chlorophyll *a* spectrum, but it narrows it, similarly to Chl *b*. The algae that contain this pigment are bluish-green (*Cyanophyceae*). They are primitive organisms, intermediates between true plants and bacteria. Like bacteria, the blue-green algae do not have photosynthetic organelles bounded by a membrane (chloroplasts); like green bacteria, they contain a pigment that absorbs at 750 nm. It can be surmised that primitive plants of the type of blue-green algae have later developed in two directions—one in which the phycobilins have disappeared, and one in which one particular form of phycobilin, the phycoerythrin, has accumulated. Plants of the first type have taken over land and the surface layers of the ocean, while plants of the second kind have concentrated in the depths of the ocean. However, the division is not sharp, with some species of one kind invading the domain of the other.

CAROTENOIDS

Structure

The third type of pigments, present in all photosynthesizing cells, the carotenoids, contain an open-chain conjugated double bond system of so-called polyene type (Fig. 9.6), ending with so-called "ionone" rings. Carotenoids are either hydrocarbons, in which case they are called carotenes, or oxygen-containing compounds, in which case they are called carotenols (or xanthophylls). The three common carotenes are α, β, and γ carotene. They are "stereoisomers" having the same molecular formula $C_{40}H_{56}$ (Fig. 9.6), and distinguished only by the arrangement of their molecules in space. The carotenols have their oxygen in the form of hydroxyl, carbonyl, or carboxyl groups attached to the "ionone" rings.

Absorption Spectra

The carotenoids generally are *yellow* or orange; they have a group of two or three absorption bands located in the blue-violet part of the spectrum (Fig. 9.4*b*). For example, β-carotene, dissolved in hexane, has absorption bands at 430, 450 and at 480 nm. Certain algae—diatoms (*Bacillariophyceae*) and brown algae (*Phaeophyceae,* from *phaeos* = brown)—contain abundantly a special carotenoid, called *fucoxanthol.* In the form in which it is present in the cells, fucoxanthol has a much broader absorption band than other carotenoids; it absorbs not only in the blue but also in the green part of the spectrum. Instead of falling

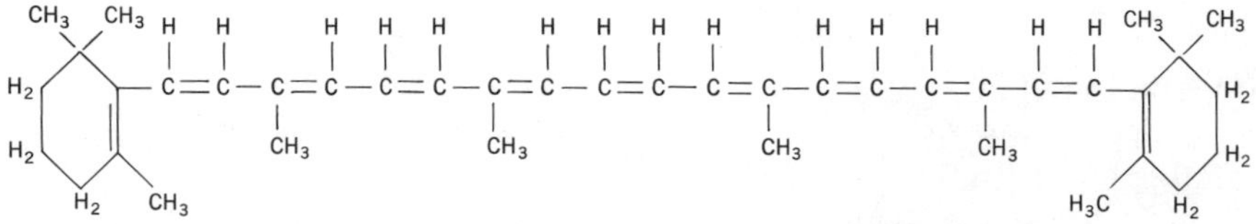

FIG. 9.6 Structure of β carotene (α and γ forms are stereoisomers). (All corners on the rings at the two ends are occupied by carbon atoms.)

off sharply to zero at about 500 nm, as in Fig. 9.4*b*, the blue absorption band of fucoxanthol declines slowly, covering much of the gap left by chlorophyll in the green, and destroying the pure green color of the plant.

To sum up, plants have developed, for the needs of photosynthesis, not a single pigment, but various pigment assortments. These change from one phylum of plants to another. The relative amounts of the several components change not only from phylum to phylum, or species to species, but also from specimen to specimen—from a shade-adapted to a sun-adapted leaf on the same tree, from a young to an old algal cell, from a specimen grown in green light to a specimen grown in red light, etc. The varying ratios of different pigments in the living cell reflect a dynamic balance of continuous synthesis and decomposition of the pigments.

It has been long a subject of guessing why chloroplast pigments occur in pairs, such as chlorophyll *a* and chlorophyll *b*, phycoerythrin and phycocyanin (a little phycocyanin is always found in red algae), and carotenes and carotenols. One member of the pair is slightly more oxidized than the other. So far, no explanation exists.

MULTIPLICITY OF CHLOROPHYLL *a* FORMS IN VIVO

Looking at the red absorption band of chlorophyll *a* in the living cell (Fig. 8.12*c*), one notes its greater width compared to the corresponding band in solution. (The "half-width" of the band—that is, its width at the level corresponding to half its maximum intensity—is 30 nm in vivo, as against about 18 nm in organic solvents.) Recently, careful observations suggested that the width of the red band in vivo is due to the presence of at least two and probably three (or even four) components.

The analysis of the absorption bands of chlorophyll *a* in suspensions of chloroplasts, or in whole algal cells, encounters several difficulties.

1. Other pigments besides chlorophyll *a* are present; this makes, in particular, the analysis of the Soret band almost impossible.
2. Light is strongly scattered by the suspension. There is general

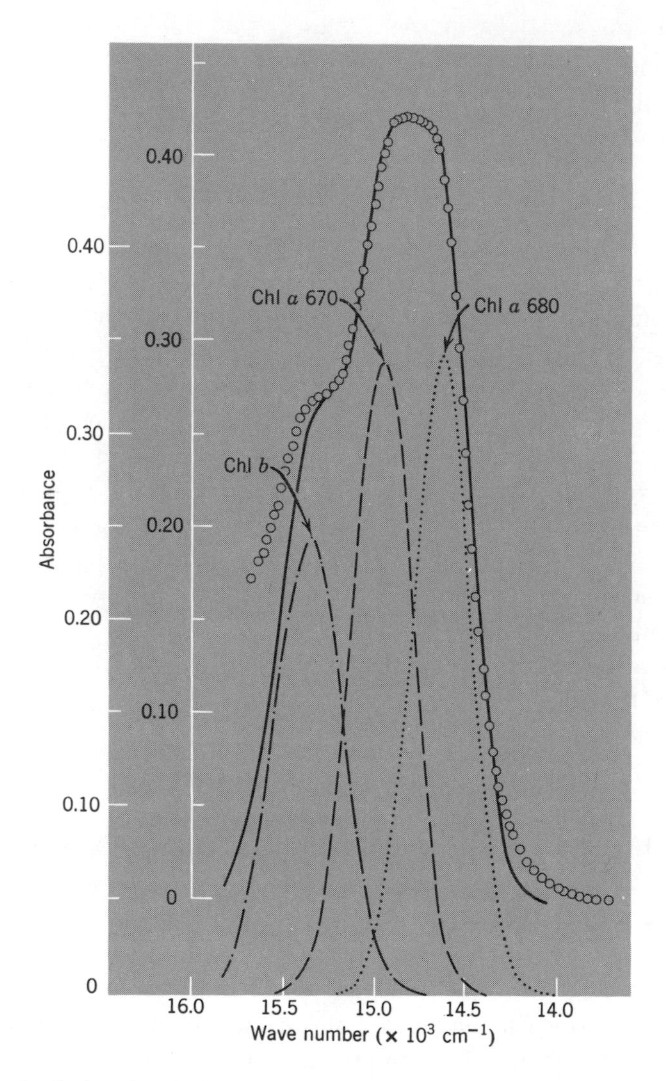

FIG. 9.7 Analysis of red chlorophyll absorption band in *Chlorella pyrenoidosa,* into two major Chl *a* components (Chl *a* 670 and Chl *a* 680), and one Chl *b* component (Chl *b* 650). Minor components of Chl *a* are supposed to exist at 690–700 nm. (C. Cederstrand, E. Rabinowitch, and Govindjee, 1966.) This figure is plotted on a wavenumber (reciprocal wavelength) scale.

scattering and, more importantly, there is "selective scattering" around the absorption band, found in 1956 by P. Latimer in our laboratory. Scattering increases the apparent absorption, if measured in an ordinary spectrophotometer, and can totally distort the absorption spectra.

3. In a scattering suspension, light that is less strongly absorbed has a larger effective path length than the more strongly absorbed light (this is known as the *detour effect*).

4. A mutual shading of the pigment molecules takes place in densely colored particles, such as the chloroplasts, while some light passes the suspension without hitting any particle at all (this has been called the *sieve effect*).

5. Fluorescence contributes to the transmitted light, particularly significantly when the absorption is strong and the fluorescence yield is high.

Two methods have been used to remedy some of these difficulties: (a) An opal glass can be placed between the sample and the detector, so that most of the forward-scattered light is collected (much of the scattered light in a suspension of large particles leaves the suspension in the forward direction). (b) An integrating sphere (Ulbricht sphere) can be used for collecting all (or, at least, almost all) the scattered energy. However, neither of the two methods eliminates the sieve or the detour effect.

C. S. French at the Carnegie Institution of Washington at Stanford, California, has constructed a "derivative" spectrophotometer in which the first derivative of the absorbance is plotted automatically as function of wavelength. This procedure accentuates the complex structure of a band, since every inflection in the band envelope appears either as crossing of the abscissa or as a peak. The "derivative absorption spectra" of various algae, obtained in this way, were analyzed by French and co-workers into three components (to which a certain standard shape, that of a so-called Gaussian error curve, has been ascribed for the purpose of analysis). They were identified as Chl *a* 672, Chl *a* 683, and Chl *a* 694. In our laboratory, Carl Cederstrand has measured the absorption spectra of algae in an integrating dodecahedron with one photocell on each of its 12 faces, and analyzed the results with the help of a computer. The red band of chlorophyll *a* was found to be essentially double with its envelope covering two neighboring bands (Fig. 9.7), like a couple huddling under a single raincoat (a couple with a child, if the chlorophyll *b* band is included). One band was located at 668 nm (French's "Chl *a* 672"), the other at 683 nm (French's "Chl *a* 683")—again assuming a Gaussian shape of the component bands. This assumption is not completely arbitrary, since a Gaussian curve was found to match very closely the red band of Chl *a* in solution. (Deviations from the Gaussian curve are, however, visible on both ends of the band in Fig. 9.7.)

Later, we found that elimination of the "sieve effect" by breaking Chlorella cells with ultrasonic waves (making the particles small enough to reduce shading to insignificance), does not affect the qualitative results of the analysis; but, as expected, it moves the peaks of the two chlorophyll *a* components closer together—to 670 and 680 nm, respectively. The half band-width of the Chl *a* 670 and Chl *a* 680 bands was found to be about 17 nm, that is very similar to that of the Chl *a* band in ether.

Cederstrand's analysis gave no evidence for the existence of a third long-wave component (such as French's Chl *a* 694). The computer, asked to explain in the simplest possible way the shape of the red band envelope in terms of Gaussian components, answered that three components (Chl *b* and two forms of chlorophyll *a*) provide as good an approximation as can be reasonably expected. Because of deviations from Gaussian shape in the wings in the Chl *a* band in vitro, it seemed unjustifiable to postulate additional components in order to obtain a better fit in the long-wave wing of the band in vivo.

However, quite apart from this analysis, there is considerable evidence for the existence of one (or two) minor absorption bands of Chl *a* in the 690–700 nm region. One is the position and the shape of the "red drop" in the action spectra of Chl *a* fluorescence and of photosynthesis (see Chapters 13 and 15). It indicates a minor Chl *a* form, not contributing to fluorescence, and contributing only partially to photosynthesis, with an absorption band in the region of 695 nm. Another is the "difference spectrum" obtained by subtraction of the spectrum of Chlorella sonicates obtained at neutral pH and under anaerobic conditions, from that of sonicates obtained at acid pH under aerobic conditions. This spectrum, too, shows a band at 693 nm, suggesting that Chl *a* 693 is present in a small amount in living cells, and is particularly easily destroyed by ultrasonic irradiation in an aerobic acid medium.

If one introduces a 693 nm band into Cederstrand's analysis, the result is a slight shift of the main long-wave component, Chl *a* 680, towards the shorter waves, and a change in the ratios of the two peaks in favor of Chl *a* 670.

Additional minor Chl *a* components (0.3% or less of total chlorophyll), with maxima at 700 nm (P700) and at 682–690 (P690) have been suggested on the basis of difference spectroscopy (Chapter 14). Their concentrations are too small for them to be noticeable in the analysis of the absorption spectrum.

Spectroscopic evidence for the complexity of chlorophyll *a* in vivo is supported by chemical evidence, presented particularly by T. M. Godnev and A. A. Shlyk in Minsk (Byelorussia). They noted that during gradual extraction of chlorophyll from leaves, the absorption spectrum changed. J. H. C. Smith and co-workers, at the Carnegie Institution in Stanford, have similarly demonstrated that during the greening of leaves, the red absorption band undergoes shifts. In these experiments, leaves were permitted to form in the dark. They are then colorless or "etiolated." When exposed to light, they rapidly become green. Altogether, it seems as if two or three spectroscopically different forms of chlorophyll *a* are formed at different times, and are extracted with different ease.

Fractionation of these two (or three) Chl *a* forms by breaking the cells mechanically and solubilizing the pigment complexes selectively by means of detergents such as desoxycholate or digitonin, or by extraction with solvents of varying polarity, has been attempted in several laboratories. These observations are relevant to the question of whether some components of Chl *a* are associated with one or the other of the two photochemical reactions postulated in photosynthesis and with the two "pigment systems" supposed to sensitize them. We shall discuss this point in Chapter 16.

Chapter 10

Absorption of Light and Fate of Excitation Energy in Plant Cells

In Chapter 8, we became acquainted with the stage on which the drama of photosynthesis is enacted. We know (from Chapter 6) that this drama is in two acts: one played, as it were, on a lighted, the other on a darkened stage. In the following chapters, we shall consider a little closer what is known about the text of these two acts. A great amount of research has been done in recent years both on the biochemical and on the photochemical (biophysical) stages of photosynthesis. The study of the dark steps has profited greatly from the discovery of the long-lived radioactive isotope, ^{14}C. Its use has facilitated the identification of the sequence of biochemical reactions that occurs in the seconds and minutes following the beginning of illumination of a photosynthetic organism (see Chapter 17).

The photochemical stage requires a different approach. In this case, we want to learn something about the nature and sequence of processes that occur in fractions of a second. We must follow rapid changes in the composition and properties of the photosynthetic apparatus; the approach must be physical rather than chemical.

The photochemical part of photosynthesis begins with light absorption. Let us review some general aspects of this process; its knowledge is essential for the understanding of the mode of action of the photosynthetic pigments.

What happens when light is absorbed by matter? The peculiar thing is that light spreads like a wave, but is absorbed like a stream of particles. The whole world picture of modern physics has this duality; everything is both wave and particle. More precisely, some aspects of the behavior of all physical objects are best described by means of the wave picture, and others, by means of the particle picture. When we talk here about physical objects, we mean primarily the so-called elementary particles of matter—electrons, protons, and neutrons, as well as particles of light, called photons or light quanta.

Heisenberg's *uncertainty principle* permits us to use both, on the face of it, mutually exclusive pictures. This principle states that every time we try to measure the magnitudes that characterize a wave, such as its wavelength or amplitude, the experiment causes the magnitudes characteristic of the particle, such as its location in space and its velocity, to become indefinite, and vice versa. Consequently, it is impossible to devise an experiment that would prove one of the two pictures correct, and the other wrong. We can, therefore, continue using one picture in describing certain aspects of a phenomenon, and the other picture in describing other aspects of it. For example, in the case of light, its propagation, reflection, and interference are best described using the wave picture, while its emission and absorption by atoms and molecules are best described by means of the particle picture.

From experiments on light *propagation* (refraction and interference) the characteristic properties of a light wave, its *wavelength,* λ, wave number, $\tilde{\nu}$ and *frequency,* ν, can be derived. They are related by the equation

$$\lambda\nu = \frac{\nu}{\tilde{\nu}} = c \qquad (10.1)$$

where c is the velocity of propagation of light (3.0×10^{10} cm/sec in vacuum). As mentioned before, the wavelengths of visible light lie, roughly, between 400 and 800 nm. Below 400 nm lies the ultraviolet, above 800 nm, the infra-red. From experiments on light absorption and emission, we derive the size of the particles, that is, energy quanta (or photons), E. They are the energy packages the light derives from matter in the process of emission, and imparts to it in the process of absorption.

A light wave hits a material object exposed to it like rain hits the pavement, in single drops, each carrying the energy E.

The fundamental relation between E and the properties of the corresponding wave is:

$$E = h\nu = hc\tilde{\nu} \tag{10.2}$$

where h is fundamental constant of physics, called Planck's constant (after Max Planck, the German physicist who first proposed the concept of an energy quantum in 1900). Its dimensions are that of action, that is, energy multiplied by time. Its value in absolute units is 6.62×10^{-27} erg sec.

Equation 10.2 establishes a connection between the wave properties $(\lambda, \tilde{\nu})$ and the particle properties of light (E). The higher the frequency—that is, the shorter the wavelength—the larger the quantum. Table 10.1 shows the wave numbers $(\tilde{\nu})$, frequencies (ν), and energy quanta, E, corresponding to certain wavelengths of the electromagnetic spectrum.

The quantum size is given in ergs, which is the absolute mechanical energy unit in the so-called CGS system; in Kcal (which is the commonly used unit in chemistry) and in electron-volts (the unit used in atomic physics). The figures 10×10^{-12} to 1.25×10^{-12} erg and 6.3 to 0.79 eV, are the energy contents of a single quantum of 200 to 1600 nm light; while 144 to 18 Kcal are the total energies of an einstein

TABLE 10.1 Optical Radiations

Wavelength, λ, (nm[a])	200	400	800	1,600
Wave number $(\tilde{\nu})$ (cm^{-1})	50,000	25,000	12,500	6,250
Frequency, ν (sec^{-1})	15×10^{14}	7.5×10^{14}	3.75×10^{14}	1.88×10^{14}
Photons (quanta), E, in erg/quantum	10×10^{-12}	5×10^{-12}	2.5×10^{-12}	1.25×10^{-12}
in Kcal/einstein[b]	144	72	36	18
in electron volts/quantum	6.3	3.15	1.57	0.79

[a] 1 nm = 10^{-7} cm [b] One einstein = 6.0×10^{23} quanta.

(6×10^{23} quanta). In Chapter 2 we gave about 112 Kcal/mole as the energy stored in photosynthesis, when one mole of CO_2 is reduced and one mole of O_2 produced. We note now that this corresponds to the energy of approximately three einsteins of 800 nm light.

PROPERTIES OF SOLAR RADIATION

The sun emits a continuous spectrum of radiation, from the ultraviolet through the visible into the infrared. It is the spectrum of a hot body having a surface temperature of 6000°C. The higher the temperature, the further the maximum of thermal radiation moves toward the shorter waves. (This is why metals appear, upon heating, first red, then yellow, and ultimately white or even bluish.)

As temperature rises and the total heat radiation increases, the *average size, E,* of quanta grows, too. At 6000°C, the maximum intensity of the emitted light lies in the orange part of the visible spectrum, at about 600 nm. According to Planck's radiation formula, it drops slowly towards the longer waves and sharply towards the shorter waves.

Only a relatively small amount of solar radiation falls into the ultraviolet, $\lambda < 400$ nm. Of the ultraviolet light present when this radiation reaches the top of the atmosphere, the larger part does not penetrate to the surface of the earth. The shorter-wave ultraviolet ($<$300 nm) is absorbed by the oxygen molecules, O_2, which are thus chemically activated, and react to form ozone molecules: $3O_2 \rightarrow 2O_3$. The ozone molecules themselves absorb in the nearer ultraviolet, between 300 and 350 nm, and react back to O_2.

Together, the two forms of oxygen form a screen that prevents most of the sun's ultraviolet light, except that above 350 nm, from reaching the surface of the earth. This is a very important protection. All organic molecules absorb light in the far ultraviolet; all living matter on land would have been destroyed if organisms were continuously exposed to unscreened sunlight. Only aquatic organisms could survive under a sufficiently thick layer of water.

About one half of the solar energy flux, as it reaches the surface of the earth, consists of visible light, that is, light with wavelengths from 400 to 800 nm. The other half is infrared light, with wavelengths above 800 mm. Some bands in the infrared region are absorbed in the air by water vapor and carbon dioxide.

Since, in photosynthesis, light has to be used to induce a chemical change, long-wave infrared light is of no use for it. In order to cause a chemical change, the quantum must be large enough to affect significantly the chemical stability of a molecule, that is, it must not be too small compared to the energies of the chemical bonds.

The quantum does not need to be so large as to break a chemical bond and decompose the molecule into free atoms or radicals; usually, it is enough for it to "energize" the molecule, transferring it into a sufficiently energy-rich *excited state* to surmount the "activation barrier" and permit a chemical reaction. (A chance of immediate remarriage makes divorce from the original partner easier!)

Excitation of a stable molecule is, however, unlikely to be chemically effective unless the absorbed quantum weakens a chemical bond *substantially*. The energies of stable chemical bonds in organic molecules are, as stated in Chapter 5, of the order of 50–100 Kcal/mole. Light quanta must be, say, half, as large, 25–50 Kcal/einstein, to weaken them significantly.

Coming from the infrared side, light quanta reach 25 Kcal/mole (or about 1 ev) at 1200 nm. They reach 50 Kcal/mole, or about 2 ev, in the middle of the visible spectrum. In other words, ultraviolet light and short-wave visible (violet, blue, and green) light, can be expected to be photochemically fully active. Certain photochemical reactions are possible also in the long-wave visible light (yellow and red) and even in the near-infrared. Quanta belonging to the long-wave infrared can cause only slight perturbations of stable chemical bonds, insufficient to lead to photochemical reactions.

Since vision itself is based on a photochemical action of light, its long-wave limit, (that is, the limit between infrared and visible spectrum), coincides approximately with the above-suggested limit of photochemical activity. The limit of visibility toward the shorter waves is imposed by the absorption of ultraviolet light in the tissues separating the retina from the air. This made it useless for animals to equip their retinas with photosensitive compounds absorbing further in the ultraviolet. The human eye can see up to about 760–800 nm on one end of the spectrum, and down to about 360–400 nm on the other hand.

Bacterial photosynthesis occurs mainly in the near-infrared, up to

1000–1100 nm. Properly sensitized photographic plates respond to light up to 1500 nm.

BEER'S LAW

When a beam of photons having appropriate frequencies strikes an absorbing substance, this beam is progressively weakened as it penetrates into the medium. This weakening follows Beer's Law:

$$I = I_0 10^{-\alpha c x} \tag{10.3}$$

where I_o is the intensity of the incident beam (the number of quanta striking each second one cm^2 of the surface normal to the direction of the beam), I is the residual intensity in depth x, and αc, a constant of the material, characterizing its absorption capacity. The product αc can be separated, in the case of solutions of colored molecules in a colorless medium, into a concentration factor, c, and an intrinsic factor, α. If c is expressed in moles/liter, α is the so-called *molar absorption coefficient*. It depends on the wavelength of light. If the beam is not monochromatic, that is, if it contains a variety of photons (for example, if it is a beam of white light), its spectral composition changes as it passes through the absorbing layer. The wavelengths for which the α values are highest are eliminated first, and those for which the α values are smallest penetrate deepest, or emerge on the other side of the absorbing layer.

The dependence of I on depth x in Eq. 10.3 is exponential. Differentiation converts Eq. 10.3 into Eq. 10.4, which applies to a thin layer, dx:

$$\frac{dI}{dx} = - I_0 \alpha c \; 10^{-\alpha c x} \ln 10 \tag{10.4}$$

By using natural logarithms, ln e (logarithms to the base e), instead of decadic ones, one can get rid of the factor ln 10. This is sometimes done, but more often, the decadic absorption coefficient is used instead of the natural. (The latter is often designated as ϵ, with $\epsilon = \ln 10 \times \alpha$.)

Equation 10.4 suggests that whenever light penetrates a certain distance dx, it is weakened in the same proportion. If it is weakened by 50% in the first 1 cm layer, it is weakened again by 50% in the next

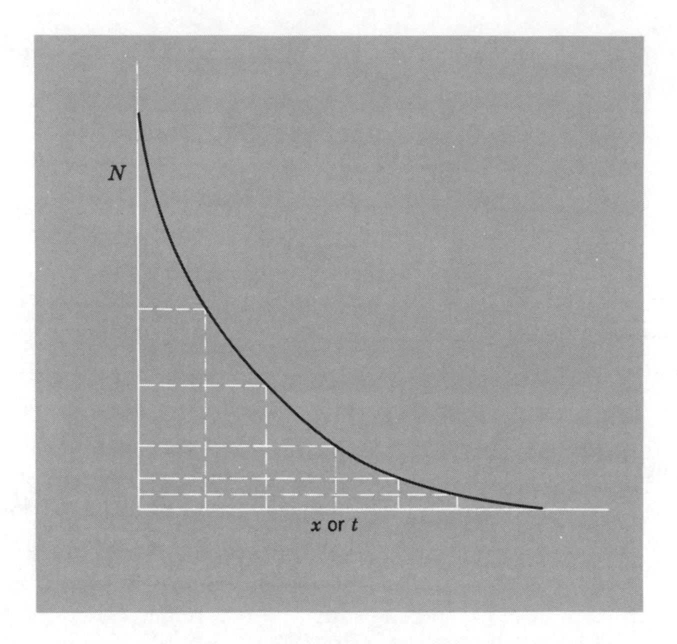

FIG. 10.1 Exponential decay curve.

1 cm layer, and so on. This is called exponential decay, a law of decay characteristic of all processes consisting of individual events not related to each other. (In our case, a molecule picks up a quantum without relation to what happened to any other quantum or any other molecule.) This is a so-called *first order process*. The number of absorption acts occurring in each layer of thickness, dx, is, under these conditions, proportional to the intensity of the beam incident on it.

Similar relations prevail in "first order" chemical reactions and in the transformation of radioactive elements; only there, the time coordinate, t, takes the place of the space coordinate x. The number of reacting molecules (or of nuclear transformations), dN, in each time element, dt, is proportional to the number of unchanged reactants (or radioactive atoms) present at the beginning of this period (Fig. 10.1). The decay is logarithmic *in time;* it is logarithmic *in space* in the case of light absorption.

One can rewrite Eq. 10.3 in the form:

$$\log \frac{I_0}{I} = \alpha c x \tag{10.5}$$

The expression log (I_0/I) is often called *optical density* or *"absorbance"* of the absorbing material. Optical density 1 means that the light is weakened by a factor of 10; optical density 2, that it is weakened by a factor of 100, etc. The optical density of 1 cm thick layer is 1 if $\alpha = 1/c$. Strong organic pigments are characterized by α-values of the order of 10^4, so that an optical density of 1 is that of a 1 cm layer of an approximately 10^{-4} molar solution.

The exponential decay of light intensity upon penetration into an absorbing medium is a nuisance in quantitative photochemical work. To study the kinetics of photochemical reactions, one would like to illuminate the photosensitive system *uniformly*, but Beer's law makes this impossible. It is helpful to use an optically very thin (or dilute) system, with a low αc-value. When I is *almost* equal to I_0, the exponential can be expanded into a series, and the series broken off after the first term, giving:

$$I = I_0(1 - \alpha c x \ln 10) \tag{10.6}$$

Exponential decay is thus replaced by *linear* decay. This approximation is permissible as long as the term $\alpha c x$ is large compared to the next, neglected term in the series, $(\alpha c x)^2/2$. If the optical density $\alpha c x$ is 0.1, the neglected term is 0.005, so that neglecting it means an error of 5 percent. If the optical density is 0.01, the neglected term is 0.00005; this means an error of 0.5%. If errors up to 1 percent can be tolerated, linear approximation can be used for samples with absorbances of the order of 0.02.

In working with photosynthesizing plant cells, one has to contend with the fact that a single chloroplast may absorb 50 or 60 percent of incident light in the peak of the absorption band (675 nm). It is, therefore, impossible to achieve uniform illumination of all chlorophyll molecules in a single chloroplast, except by using monochromatic light far from the peak of the absorption band of chlorophyll.

MECHANISM OF LIGHT ABSORPTION AND EMISSION

What happens to an atom or molecule that absorbs a photon? When light is absorbed by matter, each quantum is taken up by a single atom or molecule and the whole energy of the quantum is communicated to it. The absorbing atom or molecule is thus *excited,* that is, lifted from its normal state of lowest energy (highest stability) to an *excited,* energy-rich state. According to Bohr's theory of atomic and molecular structure, an atom or molecule can exist only in a series of discrete states of electronic energy. This is represented in Fig. 10.2, in which

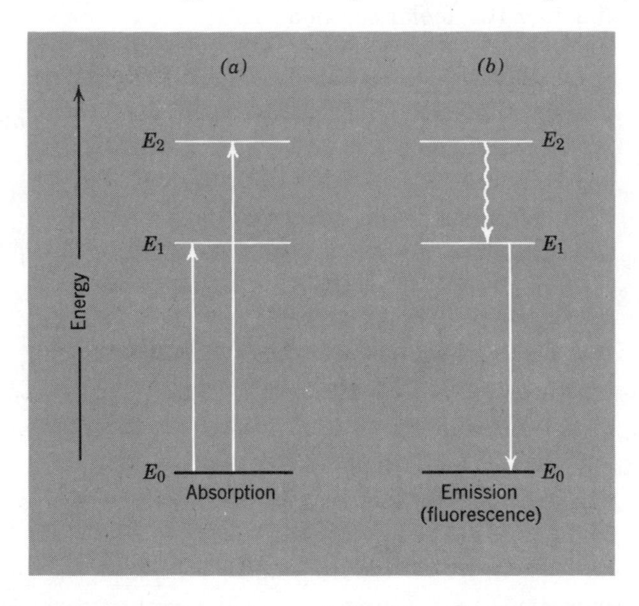

FIG. 10.2. Ground state (E_0) and two excited states (E_1, E_2) of a molecule (vibrational and rotational levels not shown).

energy levels are indicated, as they usually are in spectroscopy, by properly spaced horizontal lines. The lowest line represents the ground state, E_0, of the atom or molecule in which it exists in the absence of external activation. The higher lines (E_1 and E_2) represent *excited electronic* states—the only kind of excited states possible in an atom.

In a diatomic or polyatomic *molecule,* one or several series of (also quantized) *vibrational* and *rotational* states are superimposed on each electronic state. If the molecule is complex, the various vibrational and rotational states lie so close together that sharp absorption and emission lines of atoms, or structured bands of simple molecules, are replaced by broad, continuous bands.

The absorption lines (or bands) are represented, in schemes of this type, by arrows directed upwards (Fig. 10.2*a*) and the emission lines or bands by arrows directed downwards (Fig. 10.2*b*). The energy of the quanta of emission or absorption are proportional to the lengths of the arrows. An atom or molecule can absorb only energy quanta corresponding to the distances between the permitted energy states. Once it is excited, say to state E_2, it can emit only light quanta represented by downward arrows, leading from E_2 to a lower energy state, and finally to E_0 (Fig. 10.2*b*).[1] In *fluorescence,* light absorption leading, say, from E_0 to E_1 is reversed by light emission leading from E_1 to E_0.

When a photon is absorbed, the molecule usually is not merely transferred into an excited electronic state, but also acquires some vibrational energy. According to the so-called *Franck-Condon principle,* the absorption of a photon is a practically instantaneous process, since it involves only the rearrangement of practically inertia-free electrons. The much heavier atomic nuclei have no time to readjust themselves during the absorption act, but have to do it after it is over, and this readjustment brings them into vibrations. This is best illustrated by potential energy diagrams, such as that in Fig. 10.3. It is an expanded energy level diagram, with the abscissa acquiring the meaning of distance between the nuclei, r for example, distance x-y in a diatomic molecule xy). The two potential curves show the potential energy of the molecule as function of this distance for two electronic states, a ground state and an excited state. Excitation is represented, according to the Franck-Condon principle, by a *vertical* arrow (A). This arrow hits the upper curve, except for very special cases, not in its lowest point, corresponding to a nonvibrating state, but somewhere higher. This means that the molecule finds itself, after the absorption act, in a nonequilibrium state and begins to vibrate like a spring. This vibration is described, in Fig. 10.3, by the molecule running down, up, down again, etc., along the upper potential curve, like a pendulum. The periods of these vibrations are of the order of 10^{-13} or 10^{-12} seconds. Since the lifetimes of excited electronic states are of the order of 10^{-9} sec (or longer), there is enough time during the excitation period for many thousands of vibrations. During this time much if not all of the extra vibrational energy

[1] The wavy arrow in Fig. 10.2*b* relates to another, radiationless way in which a transition can occur—by energy loss to surrounding molecules, or by its "internal" conversion into vibrational energy of the excited molecule.

can be lost by energy exchange (temperature equalization) with the medium. The molecule, while it remains extremely "hot" as far as its electronic state is concerned, thus acquires the ambient "vibrational temperature." Fluorescence, when it comes, originates from near the bottom of the upper potential curve, and follows a vertical arrow down (F), until it strikes the lower potential curve. Again, it does not hit it in its deepest point, so that some excitation energy becomes converted into vibrational energy. The cycle absorption-emission thus contains two periods of energy dissipation. Because of this, the fluorescence arrow (F) is always shorter (that is, the fluorescence frequency is lower) than that of absorption (A). In other words, the wavelengths of the fluorescence band are longer than those of the absorption band. This displacement of fluorescence bands towards the longer waves compared to the absorption bands ("Stokes' shift") was a long-established experimental

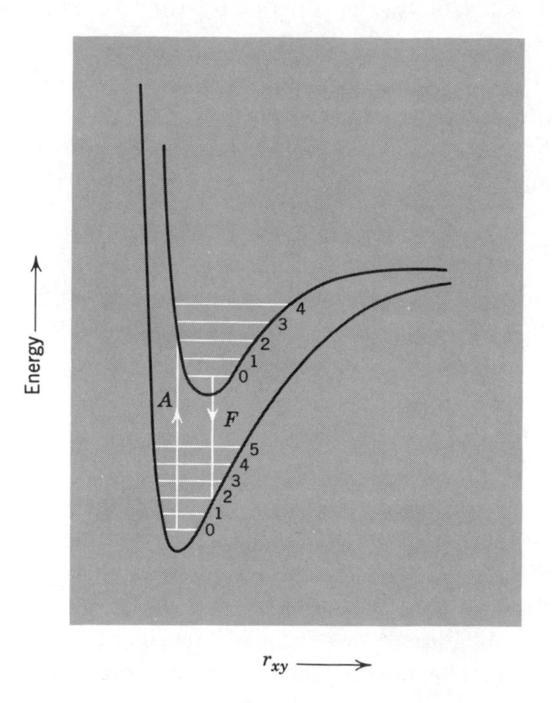

FIG. 10.3 Potential energy curves for the ground state and an excited state of a diatomic molecule. (r, interatomic distance; A, absorption; F, fluorescence; numbers indicate vibrational states.)

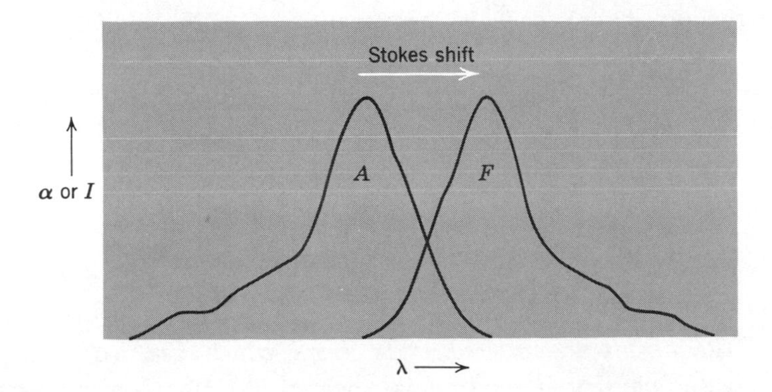

FIG. 10.4 The Stokes' shift (displacement of fluorescence band compared to the absorption band of a molecule). Approximate mirror symmetry of the two bands exists when the shapes of the potential curves in the ground state and the excited state are similar.

fact (Fig. 10.4), before the Franck-Condon principle provided its interpretation. Obviously, the extent of the shift depends on the difference between the two potential curves. The same difference determines the width of the absorption band. Chlorophyll *a*, the major pigment of all green plants, has (in organic solvents) a very narrow absorption band (half-band width, 18 nm) and a very small Stokes' shift (of the order of 7–10 nm).

FATE OF EXCITATION ENERGY IN PHOTOSYNTHESIZING CELLS

We must now consider the fate of the energy quantum absorbed by a molecule—for example, a chlorophyll molecule in a chloroplast—and the way in which it can be utilized for a photochemical reaction. A molecule, M^*, in an electronically excited state, faces several possibilities. If it is well protected from interaction with other molecules, it will have to get rid of its excitation energy "all by itself." The simplest way is for it to emit a photon back into space. This is *fluorescence*. The average time a molecule has to spend in the excited state between absorption and fluorescence is known as the *natural lifetime* τ_o of the excited state. Its duration depends on the electronic properties of the two states. An electronic charge distribution that favors the *absorption* of a quantum (that is, makes the absorption strong) also favors its *emission* (makes the lifetime of the excited state short). (A heavy earner is a heavy spender, or easy come, easy go!) This reciprocal relation permits the calculation of τ_o, the natural lifetime of the excited state, from the intensity of the absorption band leading to it. The exact relationship is rather complex,[2] we shall not discuss it here but simply mention that τ_o of the excited state from which the red absorption band of chlorophyll *a* is emitted is about 1.5×10^{-8} sec (15 nsec).

The excited molecule has, however, very rarely the chance to live out its whole natural lifetime because other ways to lose its excitation energy are offered to it (Fig. 10.5). One is intrinsic: instead of by fluorescence, the molecule can lose its electronic excitation energy by so-called *internal conversion*, that is, by conversion of the large electronic excitation quantum into smaller vibrational quanta within the molecule itself (see Figs. 10.2 and 10.5). Very often, the loss of only one or a few vibration quanta suffices to transfer the molecule from its original excited state into a nonfluorescent, *metastable* state. This is a peculiar excited electronic state, with a much longer natural lifetime. This "metastability" is due to the fact that the emission of a light quantum, transferring the molecule from the metastable state into the ground state, is extremely improbable (because it would violate certain "conservation rules"). For reasons of reciprocity, whatever is unlikely in emission is also unlikely in absorption, so that metastability of an excited state also implies weakness of the absorption band leading to it.

Usually, the ground state of a valence-saturated organic compound, that is, of a molecule having only paired electrons in the outermost

[2] An approximate relationship is:

$$\frac{1}{\tau_o} = 3 \times 10^{-9} \left(\frac{1}{\lambda}\right)^2 \Delta \left(\frac{1}{\lambda}\right) \epsilon_{max}$$

where $\frac{1}{\lambda}$ = wave number (in cm^{-1}) at the peak of the absorption band; ϵ_{max}, the corresponding absorption coefficient, and $\Delta \left(\frac{1}{\lambda}\right)$, the half-band width in wave numbers (cm^{-1}).

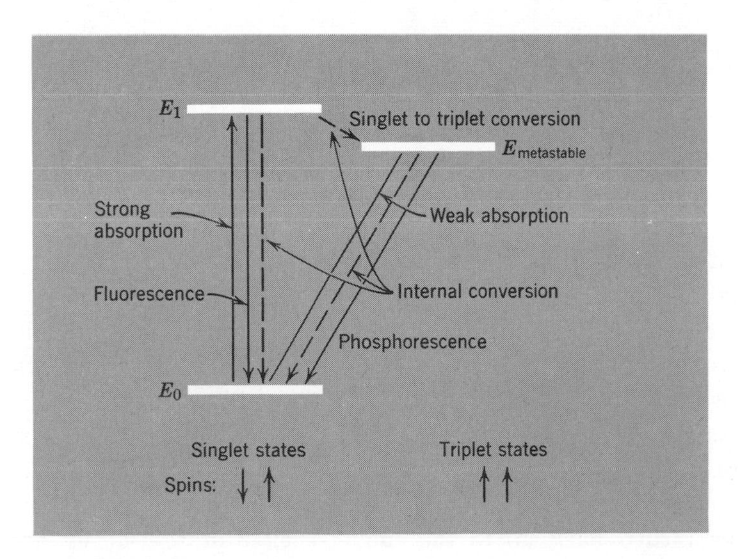

FIG. 10.5 Excitation and deexcitation processes in a valence-saturated molecule (with a singlet ground state and a metastable triplet state). (Vibrational and rotational levels are not shown.)

orbit, is a *singlet* state with an electron spin of zero, while its most important metastable state is a *triplet* state, with the net electron spin of 1, in quantum units $\lambda/2\pi$ (as indicated in Fig. 10.5).

Electrons are supposed to spin (that is, rotate around an axis passing through the electron). A triplet state is characterized by two parallel spins (↑↑): both valence electrons spin in the same sense; a singlet state, by antiparallel spins (↑↓): one electron spins clockwise, the other anticlockwise. A triplet state is so called because, upon closer examination, it proves to consist of three close components (three closely-spaced energy levels). Transition from a triplet to a singlet state or vice versa by emission or absorption of a photon is prohibited because the photon has no "handle" to accept (or to supply) the rotational momentum needed to compensate the change in rotational momentum associated with the "flipping" of one spin from parallel to antiparallel orientation in respect to the other spin.[3] The prohibited transition may, nevertheless, take place if some other rotating system (in the same molecule, or in its surroundings) offers itself to maintain the law of conservation of

[3] Conservation of rotational momentum is a basic law of mechanics.

angular momentum. Because of such interactions, a molecule in an excited triplet state will ultimately return to the singlet ground state—but only after a lifetime much longer than that of an excited singlet state. The corresponding weak (but long-lasting) emission is called "phosphorescence" (see Fig. 10.5). For the same reason, a molecule in the singlet ground state has a finite, even if a very small chance to absorb a quantum transferring it into an excited triplet state. The energy content of the quantum of phosphorescence usually is smaller than that of fluorescence because the lowest metastable level of most organic molecules is lower than the first excited singlet level. The molecule may survive in the metastable excited state for milliseconds, or even longer, dispersing its still substantial electronic excitation energy at leisure, as it were, either by continued internal conversion, or by interaction with other molecules in the medium. The electron in the metastable state may be kicked back up to the singlet-excited state (E) by acquiring some energy from the medium; fluorescence following such an action will be delayed (*delayed fluorescence*). It seems that many photochemical reactions occur through the intermediary of metastable states. These states give the light-excited molecule a better chance to encounter a reaction partner than do the fleeting fluorescent singlet states. The existence of triplet states has been proven in chlorophyll dissolved in organic solvents, but not yet clearly demonstrated in vivo; there are, however, various indications that the triplet state of chlorophyll may be involved in photosynthesis.

Fluorescence, internal conversion, and transition to the triplet state are competing spontaneous (or "monomolecular") processes, by which excited states can lose their excitation energy. In addition, there can be "induced" deactivation, as well as photochemical reactions, caused by interaction of the excited molecule with different collision partners.

In a condensed (that is, solid or liquid) system, where an excited molecule is crowded by other molecules, it is particularly strongly threatened by accidents, which can deprive it of its excitation energy. One can say that, in addition to the two inescapable "first order" death causes, the "heart disease" of fluorescence and the "cancer" of internal conversion, the life of an excited molecule in a condensed medium is threatened also by a variety of "second order" processes—communicable diseases or injuries caused by encounters with other molecules.

Measurements have shown that the fluorescence yield of chlorophyll *a* in solution (for example, in diethyl ether, or ethanol) is about 30 percent. This means that out of 100 excited molecules, 30 die by emitting a quantum of fluorescence and 70 by internal conversion (presumably initiated by transfer into the metastable state).

In the living cell, the fluorescence yield of chlorophyll *a* is much lower, In weak light, it does not exceed 3 percent, that is, at best, only 3 out of 100 excited molecules are deactivated by fluorescence. What of the others? In low light, when photosynthesis proceeds with its maximum quantum yield, they may be all, or practically all, utilized for the primary photochemical reaction of photosynthesis by interaction of the excited molecule (either before or after its conversion into the metastable stage) with a "primary reactant."

Photochemical utilization of excitation energy will or will not affect the quantum yield of fluorescence, depending on whether it precedes or follows the conversion into the metastable state. In the first case, photochemical sensitization is in direct competition with fluorescence; in the second, it is not.

According to Bohr's theory, excitation of an atom (or molecule) means the shifting of one of its electrons from its closest permitted orbit around the nucleus (energy E_0), into a wider one (energy E_1). The primary photochemical process in photosynthesis was described, in Chapter 5 as an *oxidation-reduction* (that is, in essence, shifting of an electron from the reductant to the oxidant). Photosynthesis thus requires the replacement of electron transfer from one orbit to another within a single molecule of Chl *a*, by electron transfer from one molecule (reductant) to another (oxidant). In principle, this can be imagined to occur in three steps, as represented by the three arrows in Fig. 10.6. The vertical arrow represents the excitation of an electron in pigment *P* to a higher energy level. (This must be, specifically, the pigment molecule in or near the "reaction center.") The upper slanted arrow represents the transfer of this high-energy electron to the oxidant, and the lower downward-slanted arrow, the replacement of this electron in *P* by an electron supplied by the reductant. The two dashed arrows are slanted downward, because spontaneous dark reactions must occur with some loss of free energy. The net result is oxidation of the reductant *R* and reduction of the oxidant *O*, with the energy of light utilized to move the electron "uphill" from its more stable association with *R* to a less stable association with *O*.

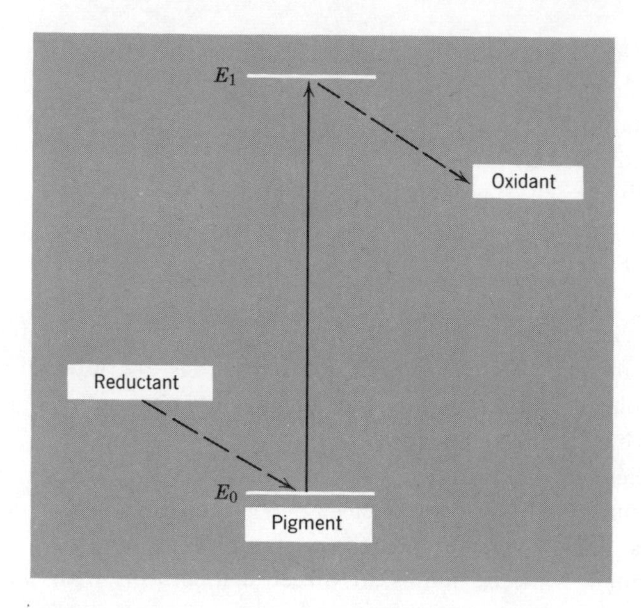

FIG. 10.6 Pigment-sensitized electron transfer from reductant to oxidant, with storage of energy. (E_0, ground state; E_1, excited state.)

In practice, of course, the process is more complicated. To begin with, it seems to involve a *transfer of energy quanta* from pigment molecule to pigment molecule in the "photosynthetic unit" (see Chapters 6 and 8) and their *trapping* in a reaction center. Recently we have begun to acquire some understanding of this process (Chapters 11 and 12). Furthermore, it seems to require two photochemical reactions in succession, with an enzymatic reaction sequence between them (Chapters 14–16).

Chapter 11

Action Spectrum and Quantum Yield of Photosynthesis

It was suggested in Chapter 9 that the function of certain accessory pigments (for example, phycoerythrin in red algae, and fucoxanthol in the brown ones), is to permit the utilization for photosynthesis of light in the middle part of the visible spectrum, not absorbed efficiently enough by chlorophyll *a*.

In 1883–85, Th. W. Engelmann, a German physiologist, placed cultures of red, brown, and green algae on the microscope stage located in the focal plane of a spectroscope. Using motile bacteria as oxygen indicators, he found that red algae produced most oxygen in the green part of the spectrum, brown algae in the blue-green part of it, and green algae in the blue-violet and the red region. Every time, the peak of photosynthetic activity coincided with the region of maximum absorption. Engelmann then offered a "law," $E_{\text{abs}} = E_{\text{photosyn}}$, suggesting that all light *absorbed* by chloroplast pigments is also light *used* for photosynthesis.

Engelmann's interesting results were forgotten when Richard Willstätter and Arthur Stoll in Munich carried out their famous investigations of chlorophyll and photosynthesis, published in two volumes in 1913 and 1917, respectively. Using green leaves exposed to incandescent light, they concluded, rather incidentally, that light absorption by carotenoids did *not* contribute to photosynthesis. This erroneous conclusion may have been due to several reasons, beginning simply with the extreme weakness of blue-violet rays in the light of an incandescent bulb. However, the authority of Willstätter (who was awarded the Nobel Prize in 1915), and general concentration on the study of chlorophyll as "the" photosynthetic pigment, made most students of photosynthesis reluctant to admit a possible contribution to photosynthesis of pigments other than chlorophyll. When a German plant physiologist, C. Montfort, concluded, from crude measurements, that light absorption by fucoxanthol contributes significantly to the photosynthesis of brown algae, little if any attention was paid to his claims.

Since 1940, however, more reliable measurements, first by W. M. Manning, H. J. Dutton, and co-workers at the University of Wisconsin, and then by Robert Emerson and co-workers (first in California and then at the University of Illinois), brought general confirmation of qualitative validity of Engelmann's law. In fact, only light quanta absorbed by the pigments dissolved in the cell sap or bound in cell walls (such as the anthocyanins) are totally lost for photosynthesis. All quanta absorbed by pigments located in the chloroplasts (which includes the carotenoids, the phycobilins, and the chlorophylls) contribute, to some extent, to photosynthesis. However, the effectiveness of this contribution varies between 20 and 100 percent of that of the quanta absorbed by chlorophyll *a* (or, more exactly, by the most effective of the several forms of chlorophyll *a*).

The question naturally poses itself: do accessory pigments carry out photosynthesis, as it were, on their own? If this were so, it would be difficult to understand why all plants capable of photosynthesis contain chlorophyll *a*, including the deep-sea red algae, in which chlorophyll *a* has little chance to absorb directly a significant number of light quanta. An alternative interpretation is offered by the observations of *sensitized fluorescence* of chlorophyll *a* in vivo (to be described in Chapter 12): it is that quanta taken up by accessory pigments are transferred by a so-called resonance mechanism to chlorophyll *a*, and contribute to photosynthesis only after this transfer. They act as if they were servants who keep their master, chlorophyll *a*, well supplied with quanta, so he can keep up his photocatalytic trade.

It is thus legitimate to postulate that one function of the accessory pigments in photosynthesis is to supply quanta to chlorophyll *a*. In the case of yellow carotenoids, where the efficiency of this transfer is low, it is logical to surmise that these pigments have also another and more significant function, either in photosynthesis or in some other metabolic process. What that function is, we do not yet know.

THE ACTION SPECTRA OF PHOTOSYNTHESIS

We shall now consider the problem more quantitatively. Light of a single wavelength, λ (or a single frequency, ν) is called *monochromatic*. (In practice, "monochromatic" light always covers a finite band of wavelengths.) In photochemical research, it is often desirable to measure the chemical effect produced by monochromatic light. This requires the use of powerful light sources, and of monochromators of high resolving power. The term "action spectrum" is used to describe a plot of the intensity of some phenomenon (for example, the rate of photosynthesis) produced by monochromatic light as function of the wavelength (or frequency) of this light. More specifically, one measures the rate of O_2 evolution (or of CO_2 uptake), P, produced by a known number of *incident* quanta, I_i—or, better, by a certain number of *absorbed* quanta, I_a—of a given wavelength. Both plots:

$$\frac{P}{I_i} = f(\lambda) \tag{11.1}$$

and

$$\frac{P}{I_a} = f(\lambda) \tag{11.2}$$

have occasionally been described as "action spectra." We shall use this term for the more significant plot (11.2).

If P is expressed as the number of O_2 molecules evolved per second, and I_a as the number of quanta absorbed per second, P/I_a is the *quantum yield* (Φ) of photosynthesis. If all pigments that absorb light were equally effective in photosynthesis, in all parts of their absorption bands, the plot of Eq. 11.2 would be a horizontal line. It is the deviation from the horizontal that is of interest.

Light of a certain wavelength is often absorbed by several pigments present in the cell. If the *action spectrum* of photosynthesis, obtained by systematic measurements of P in monochromatic light throughout the visible spectrum, is compared with the *absorption curves* of the several pigments present, conclusions can be drawn as to the relative efficiency with which different pigments contribute to photosynthesis. Such measurements require certain precautions. We know that the pigments participate in the primary photochemical reaction, which determines the rate of photosynthesis in *weak* light; while in *strong* light (in the "light-saturated state"), the overall rate is determined by a dark enzymatic reaction. To give significant results, action spectra must be measured in the light-limited state, that is, at low light intensities. In strong, saturating light, all the revealing features may be smoothed out.

One practical problem in photochemistry is to obtain sufficient light energy (that is, a sufficient flux of light quanta) within a narrow "monochromatic" band to produce a measurable chemical effect. When the effect of a single light quantum can be multiplied thousands of times, as in the development of a photographic plate, the problem is relatively easy; but when Einstein's law applies, that is, when one absorbed quantum changes only one molecule, energy limitations become bothersome. Furthermore, in radiation from the incandescent lamp, the intensity drops rapidly from the red to the blue end of the spectrum. This is why we have much better data on the dependence of the yield of photosynthesis on wavelength in the long-wave, orange-red region, than in the short-wave, blue-violet spectral region. (Xenon lamps may be used in the blue region, but, unfortunately, their emission is rather unsteady.)

Very crude action spectra can be obtained by means of colored glasses, or colored gelatin filters, which transmit broad spectral bands. Much better are the so-called interference filters, which isolate, from a continuous spectrum, bands about 10 nm (or even 5 nm) wide. With a set of such filters, one can explore the whole visible and near ultraviolet spectrum. Still better, however, is to use a *monochromator*—a powerful spectroscope—isolating from a continuous spectrum nearly monochromatic spectral bands.

Truly significant action spectra are plots of the *maximum quantum yield* (Φ) of photosynthesis, in monochromatic light, as function of the wavelength (or frequency) of this light. The most reliable and systematic measurements of this type have been made by Robert Emerson and co-workers, with the green alga, *Chlorella pyrenoidosa* (Fig. 11.1), the red alga, *Porphyridium cruentum* (Fig. 11.2), the blue-green alga, *Chroococcus,* and the diatom, *Navicula minima* (Fig. 11.3). These measurements were made with precision manometers and a large grating monochromator.

One fact is immediately obvious: light absorbed in the blue-violet region of the spectrum is relatively ineffective in bringing about photosynthesis. In the face of such deficiency, one can first postulate that all quanta absorbed by a certain group of pigments (such as the carotenoids), do not contribute to photosynthesis, or contribute to it with low efficiency. In a second approximation, one has to consider the possibility that some pigments of a given group are less efficient than the others (carotenes may be less efficient than carotenols, or chlorophyll *b* less efficient than chlorophyll *a*). Finally, one has to consider the possibility that quanta of different size, absorbed by one and the same pigment (for example "blue" and "red" quanta absorbed by chlorophyll) may differ in efficiency. Forgetting for a moment this last possibility and considering action spectra only as evidence of differences in photo-

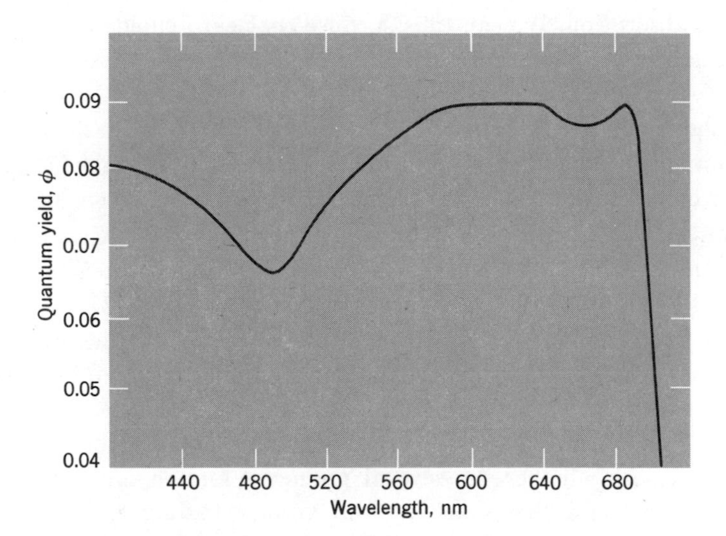

FIG. 11.1 Action spectrum of photosynthesis (quantum yield of oxygen evolution as a function of wavelength of light) in a green alga (*chlorella pyrenoidosa*). (R. Emerson and C. Lewis, 1943.)

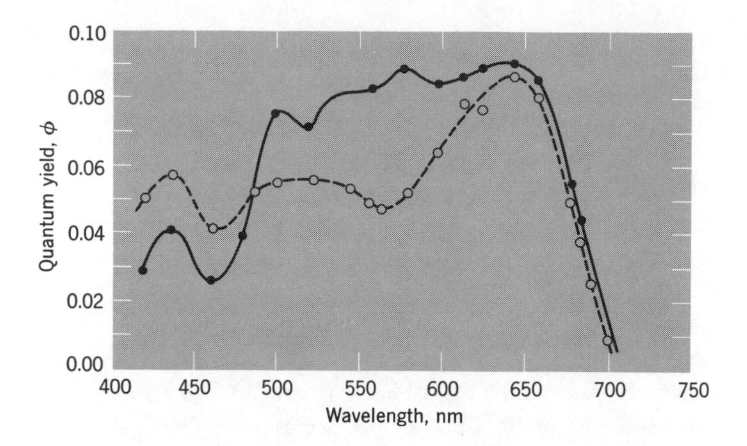

FIG. 11.2 Action spectrum of photosynthesis (quantum yield of oxygen evolution as a function of wavelength of light) in *Porphyridium cruentum* (a red alga). (Solid line–algae preilluminated with green light; dashed line–algae preilluminated with blue light.) (M. Brody and R. Emerson, 1959.)

synthetic efficiency of different pigments, we note that all action curves dip in the region (450–500 nm) where carotenoids contribute most to the total absorption. We can thus surmise that carotenoids are relatively poor sensitizers in photosynthesis.

The region in which phycobilins contribute most to absorption in red algae shows no similar dip. Neither does one appear in the region of predominant fucoxanthol absorption in diatoms (although fucoxanthol is a carotenoid), and of chlorophyll *b* absorption in green algae. This suggests that quanta absorbed by the phycobilins, by fucoxanthol, and by chlorophyll *b*, are about as efficient in photosynthesis as quanta absorbed by chlorophyll *a* (see, however, Chapters 9 and 13).

To estimate more precisely the relative efficiency of the several pigments, one has to calculate the relative number of quanta absorbed by different pigments out of monochromatic beam traversing a mixture of several of them. This is an awkward problem. In the first place, we are not sure whether the several pigments form a uniform mixture, so that none has a first crack at absorption, shading the others. If a homogeneous absorption is postulated, the allotment of quanta should be in proportion to the products, αc, of their absorption coefficients and their concentrations. The latter can be determined by extraction and quantitative analysis; but the absorption coefficients in vivo are difficult to estimate precisely. We know that upon extraction, the absorption bands of the carotenoids and the chlorophylls shift (see Chapter 9) because they are affected in vivo by aggregation of pigment molecules and by their association with molecules of the medium. Precise determinations of these shifts are not easy, and the results not very reliable because of mutual overlapping of the several absorption bands. The allocation of absorbed quanta to different pigments is, therefore, fraught with considerable uncertainties. The relative efficiencies, calculated from Emerson's data on oxygen evolution, or from Duysens' data on fluorescence (see Chapter 12) are, therefore, only more or less satisfactory approximations.

Despite these uncertainties, one can definitely assert, from the analysis of the action spectra, that differences between the several pigments are more subtle than simple "effective" or "not effective." Even the yellow carotenoids are not *entirely* ineffective; if the best sensitizing efficiency of chlorophyll *a* is set as 100%, that of the carotenoids varies, in different plants, between 20% and 50%. The effectiveness of fucoxanthol in brown algae and diatoms seems to be about 80%.

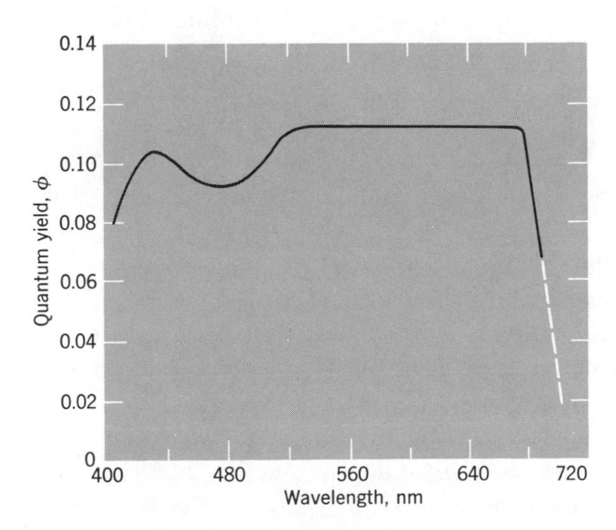

FIG. 11.3 Action spectrum of photosynthesis (quantum yield of oxygen evolution as a function of wavelength of light) in a diatom (*Navicula minima*). (T. Tanada, 1951.)

According to Marcia Brody and Robert Emerson, the relative efficiency of quanta absorbed (below 600 nm) by phycoerythrin in the red alga *Porphyridium* can vary from 1 to 0.6 of that of the quanta absorbed by chlorophyll *a* (at 650 nm), depending (see Fig. 11.2) on whether the cells had been preilluminated with bright green light (absorbed preferentially by phycoerythrin, and apparently stimulating its efficiency) or with blue-violet light (absorbed preferentially by chlorophyll *a* and apparently decreasing the effectiveness of energy transfer from phycoerythrin to chlorophyll *a*). Variations with wave-length in the efficiency of quanta absorbed in chlorophyll *a* itself, which is particularly strong in these algae, will be discussed in Chapter 13.

THE MAXIMUM QUANTUM YIELD OF PHOTOSYNTHESIS

We have dealt so far in this chapter with the *relative* quantum yields of photosynthesis in monochromatic light of different color. We found that this yield is approximately constant when light absorption occurs in chlorophyll *b* or *a* (however, see Chapter 13) and, under proper conditions, also in light absorbed by the phycobilins, but declines sharply in light absorbed by the carotenoids.

We shall now say something about the *absolute* value of the maximum quantum yield, Φ, or minimum quantum requirement, $1/\Phi$, of photosynthesis. The latter is the minimum number of absorbed quanta needed to reduce one molecule CO_2 and liberate one molecule of oxygen; the maximum quantum yield is the inverse of it, that is, the maximum fractional number of oxygen molecules produced per absorbed quantum.

From the shape of the light curve of photosynthesis (Fig. 6.1) it is clear that in order to determine the maximum yield, one has to measure the rate of photosynthesis in the "light-limited" state (as close to zero illumination as feasible). In fact, the maximum quantum yield can be defined as the initial slope of the light curve, the limiting value of $(\Delta P/\Delta I)$ as I approaches zero. In practice, measurements have to be carried out below, or only slightly above, the so-called *compensation point,* where photosynthesis just balances respiration so that the net gas exchange is zero. This means that the light intensity is such that it gives only one-tenth to one-fifth of the maximum rate of photosynthesis of which the cells are capable. (It is these quantum yields in low light that have been plotted in Figs. 11.1 to 11.3.)

A landmark in the study of photosynthesis, and one of the first applications of quantum theory to photochemistry in general, had been the first determination, by Otto Warburg and E. Negelein in Berlin in 1922, of the maximum quantum yield of photosynthesis in *Chlorella*. This study first applied the manometric method to rate measurements of photosynthesis. It also first introduced unicellular algae (*Chlorella*) as objects for such studies. The result of this investigation appeared, at first, highly satisfactory: the minimum quantum requirement of photosynthesis was found to be 4, that is, four quanta absorbed for each oxygen molecule liberated. This seemed highly satisfactory because the reduction of one molecule of CO_2 by two molecules of H_2O requires, as we have repeatedly noted, the transfer of *four* hydrogen atoms. It seemed plausible that the transfer of each atom required one quantum. Since four mole einsteins of red light (680 nm) carry (see Table 10.1) about 172/Kcal, while about 120 Kcal of free energy are stored in photosynthesis per O_2 mole liberated, Warburg's figure suggested a remarkably high efficiency, ($^{120}/_{172} \simeq 70\%$), with which plants can convert light into chemical energy.

For almost twenty years, the four-quantum mechanism of photosynthesis was generally accepted and marveled at, although it placed speculations as to the possible detailed mechanism of photosynthesis into something of a straight jacket, because of the required high efficiency. Since 1938, however, attempts to repeat Warburg and Negelein's experiments with improved techniques led to the conclusion that the true maximum quantum yield is much lower—probably, by a factor of two. Robert Emerson and his co-workers in California, and then in Illinois, have carried out these experiments most extensively by so-called differential manometry (a method also first devised by Warburg, in which two manometers with two vessels of different size are employed). It permits separate determination of ΔCO_2 and ΔO_2 (without postulating $\Delta O_2 = -\Delta CO_2$, as in the original, one-manometer method). Other investigators, particularly Farrington Daniels, W. M. Manning, and co-workers at the University of Wisconsin, obtained similar results by various other methods, including polarography (potentiometric determination of oxygen), determination of synthesized combustible material with a calorimeter, and determination of the rate of carbon dioxide consumption (by electrochemical analysis, by measurements of radioactive CO_2 uptake, of changes in the infrared absorption spectrum, or of changes in heat conductivity of the gas circulated over the plants).

In all these experiments, quantum requirements of the order of 10–12 have been usually found. The minimum values (which are the most significant ones) were close to 8. The conclusion that the normal minimum quantum requirement of photosynthesis is 8 appears compatible with the data of most researchers; scattering due to errors may account for the occasional finding of lower values. Recent (1968) precise studies on *Chlorella* sunspensions of different age, with different CO_2 concentrations, by Rajni Govindjee yielded no quantum requirements under 8. A quantum requirement of 8 suggests utilization of *two* quanta for the transfer of each hydrogen atom from H_2O to CO_2. It is significant that similar values (8–10 quanta per O_2 molecule) were found, in our laboratory, also for the Hill reaction (reduction of quinone by *Chlorella* cells, and reduction of different oxidants by chloroplast suspensions). The best available measurements of bacterial photosynthesis suggest that in this case, too, 8–10 quanta are required to reduce one CO_2 molecule, irrespective of the nature of the H-donor.

Warburg strongly opposed the findings that the minimum quantum requirement for photosynthesis is 8. Not only did he assert that his old findings, according to which four quanta are sufficient for photosynthesis, are still valid, but reported that *Chlorella* can carry out photosynthesis with a quantum requirement of only 2.8, corresponding to 100 percent efficiency in converting light energy into chemical energy! This, Warburg described as a result "he has always seen ahead," a proof that "in a perfect world, photosynthesis must be perfect."

Warburg found that minimum requirements of the order of 3 quanta per molecule O_2 could be observed for hours at a time, even in light strong enough to exceed compensation by a factor of 3 to 5.

These results contradict the findings of Emerson and co-workers, and the results obtained in other American laboratories, including a recent study by J. Bassham in Berkeley (1968). That photosynthesis can proceed with a 100 percent energy conversion yield is unlikely. It is hard to believe that a complex chemical reaction can operate, at high speed, with no "friction losses" whatsoever!

The value of 8, we believe, must be accepted at present as the most likely minimum quantum requirement of photosynthesis.

THE RED DROP

Figures 11.1 to 11.3 show one peculiarity that we have not yet considered. The action spectra drop not only at the short-wave end (where this can be attributed to the presence of carotenoids), but also at the long-wave end of the spectrum. This drop is particularly conspicuous in red algae, where it occurs at 650 nm, right in the middle of the main absorption band of chlorophyll *a* (Fig. 11.2). However, Emerson pointed out that the $\Phi = f(\lambda)$ curves in Fig. 11.1 (for *Chlorella*) and 11.3 (for *Navicula*) also show a drop, albeit only above 680 nm, that is, beyond the peak of, but still within, the chlorophyll *a* absorption band. These observations have become the starting point of an interesting development, including both experimental study and theoretical interpretation, with which we shall deal in Chapter 13.

The dip in the action spectrum of green algae in the neighborhood of 660 nm, seen in Fig. 11.1, also is reproducible and waits for an interpretation.

Chapter 12

Energy Transfer and Energy Migration in Photosynthesis

In Chapter 11, we discussed the effectiveness of different pigments in contributing energy to photosynthesis and suggested that pigments other than chlorophyll *a* cooperate by transferring their excitation energy to the latter pigment. This transfer probably takes place by a resonance mechanism, similar to the one familiar from acoustic experience, but properly describable only in terms of quantum mechanics. There are reasons to believe that most chlorophyll *a* molecules also do not participate directly in the primary photochemical process in photosynthesis, but transfer their excitation energy (by a somewhat different mechanism) to the few chlorophyll *a* molecules directly associated with the enzymatic reaction centers.

We shall deal in this chapter with energy transfer between different pigments ("heterogeneous" transfer), as well as with transfer between identical molecules ("homogeneous" transfer). The latter can be repeated many times, giving rise to energy *migration*.

Direct evidence of energy transfer between different pigments is provided by *sensitized fluorescence*. Light quanta absorbed by molecules of one pigment (for example, chlorophyll *b*) are transferred to molecules of another pigment (for example, chlorophyll *a*); when the first pigment is excited, only fluorescence of the second is observed. This phenomenon is well known from studies on gases and solutions. The occurrence of

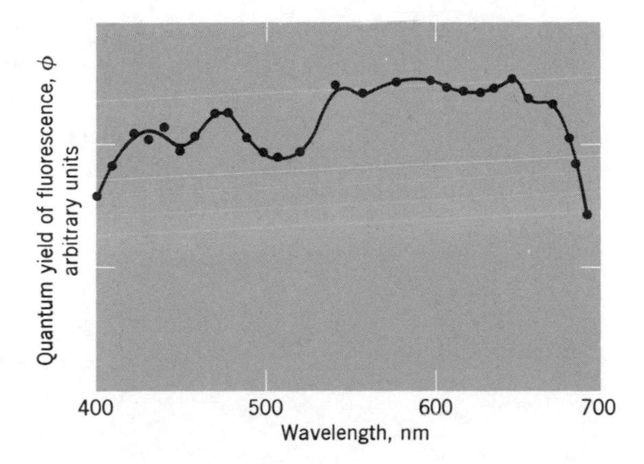

FIG. 12.1 Action spectrum of fluorescence excitation in *Chlorella pyrenoidosa;* fluorescence observed at 740 nm. (P. Williams.)

heterogeneous energy transfer between plant pigments is indicated by the action spectra (or "excitation spectra") of chlorophyll *a* fluorescence in vivo.

In Chapter 11, we described the action spectra of photosynthesis, showing that various accessory pigments contribute excitation energy to photosynthesis with different efficiency. If, instead of the yield of photosynthesis, we plot, as function of wavelength of the exciting light, the yield of chlorophyll *a* fluorescence (measured in the long-wave region where fluorescence is not reabsorbed), we obtain curves that approximately parallel the action spectra of photosynthesis (compare, for example, Fig. 12.1 with Fig. 11.1).[1] This suggests that accessory pigments sensitize photosynthesis by transferring their excitation energy to chlorophyll *a*.

In dealing with excitation energy transfer we have to distinguish between two mechanisms—the "first order" mechanism operative in "homogeneous" energy transfer (and perhaps in some cases of heterogeneous transfer as well!), and the "second order" mechanism operative in heterogeneous energy transfer.

In the first case, that of "ideal" resonance, such as exists between

[1] The reason for discrepancy between the two curves in the 400–450 nm region remains to be elucidated. Fluorescence measurements in the blue-violet excitation region can be made with much narrower bands than those of photosynthesis, thus permitting a more detailed determination of the action spectrum.

atoms of sodium in sodium vapor, or molecules of chlorophyll *a* in a photosynthetic unit, the size of quanta absorbed by one molecule equals precisely that absorbed by the others. When such identical atoms or molecules are close together, as in a dense vapor, concentrated solution, or a crystal, the interaction forces between adjoining molecules cause the excitation quantum to become a communal property of all of them—just as a free electron is a communal property of all atoms in a metal crystal. This is the basic mechanism of *homogeneous* energy transfer and migration (to be discussed later in this chapter); it could also contribute to energy transfer between different molecules if they have overlapping absorption bands.

HETEROGENEOUS ENERGY TRANSFER

The mechanism of resonance transfer of energy between unlike molecules without overlapping absorption bands was analyzed in 1948 by

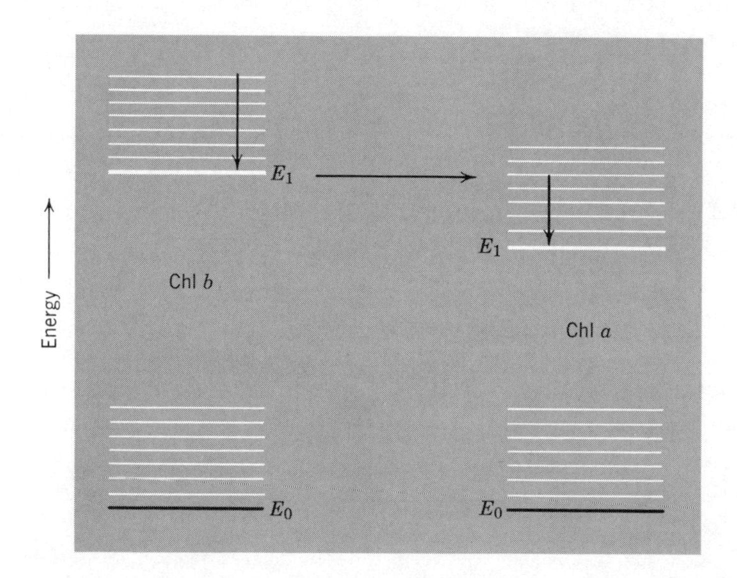

FIG. 12.2 Energy level diagrams of Chl *b* and Chl *a*, indicating why energy transfer is possible from the lowest excited state of Chl *b* to Chl *a*.

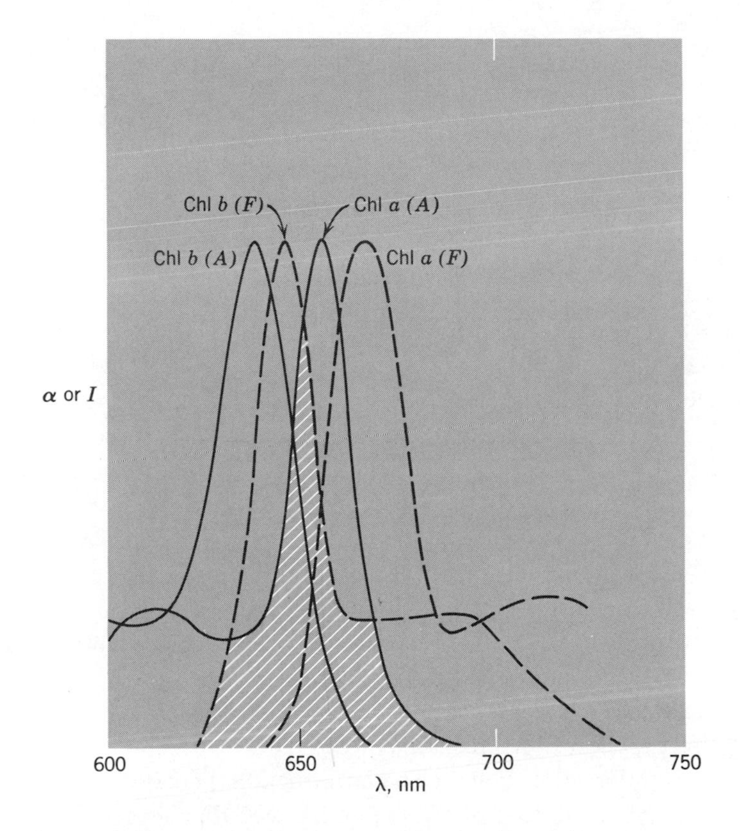

FIG. 12.3 Overlapping of absorption band (*A*) of Chl *a* with fluorescence band (*F*) of Chl *b* (shaded area), and of the absorption bands of the two pigments.

the German physicist Theodor Förster. In this case, "resonance" does not operate *during* the absorption process itself. The absorbed quantum belongs, at first, to one pigment molecule only. However, in the process of vibrational energy dissipation in the excited electronic state of the "donor" pigment (say Chl *b*), states are reached that are in resonance with certain (strongly vibrating) states of the "acceptor" chlorophyll *a* (Fig. 12-2). This "delayed" resonance is what makes energy transfer possible. According to Förster, one measure of probability of such a transfer is the overlapping of the fluorescence band of the donor and the absorption band of the acceptor, indicated by shading in Fig. 12.3. (These two figures refer to the pair Chl *b* + Chl *a* in which *both* mechanisms are possible, since the absorption band of Chl *a* overlaps both the absorption band and the fluorescence band of Chl *b*.)

The interaction between molecules with overlapping absorption bands is a dipole-dipole interaction.[2] The energy of such interaction is proportional to r^{-3}, where r is the distance between the centers of the two dipoles (presumed to be large compared to the distance between the two poles in each dipole). The interaction between molecules in which the fluorescence band of one overlaps the absorption band of the other, caused (as previously mentioned) by "delayed" resonance is, on the other hand, a "second order" effect; as such, it is proportional to r^{-6}.

The hierarchy of potential energies, E (and of interaction forces, F, which are the derivatives of potential energies in respect to distance, $F = dE/dr$), begins with the Coulomb interaction between two free charges ("monopoles"). In this case, E is proportional to r^{-1}, and F to r^{-2}. The interaction energy between a monopole and a permanent dipole is proportional to r^{-2}, and the force between them, to r^{-3}. The energy between two permanent dipoles is proportional to r^{-3} and the force between them to r^{-4}. The interactions involving an *induced* (rather than a permanent) dipole (that is, a dipole formed in a nonpolar molecule under the influence of a monopole, or of another dipole) are "second order" effects (because the moment of the induced dipole is itself proportional to the interaction force). The corresponding energy is proportional to $(r^{-2})^2$, or r^{-4} in the case of monopole-induced dipole interaction, while the interaction energy of two mutually induced dipoles is proportional to $(r^{-3})^2$ or r^{-6}. This is the type of interaction that quantum mechanics suggests exists between unlike molecules resonating in the above-described "delayed" way.

Förster's calculations suggested that, if the overlapping between the fluorescence band of an excited "donor" molecule and the absorption band of a nonexcited "acceptor" molecule is substantial (as in Fig. 12.3), the time needed for energy transfer may become equal to the natural lifetime of the excited state when the distance between two approaching molecules still is considerably larger than the sum of their "kinetic" radii (that is, radii determined by gas-kinetic methods). In other words, the probability of energy transfer can reach 50% long before the two

[2] The dipoles being the "virtual" dipoles corresponding, in quantum-mechanical treatment of light absorption and emission, to coexistence of the excited state with the ground state (no dipole corresponds to each state separately).

molecules actually collide. The calculated (and actually observed) "critical distances" over which the transfer probability equals 50%, are of the order of 5 nm. In a chloroplast, the distance between different pigment molecules is likely to be much less than 5 nm, so that the probability of energy transfer must be quite high.

Because of the Stokes shift, the fluorescence band of a pigment absorbing at the shorter waves often overlaps the absorption band of a pigment absorbing at somewhat longer waves, but not vice versa (Fig. 12.3). Consequently, energy transfer usually can go in one direction only. For example, in red algae, energy transfer goes from pigments absorbing in the blue region of the spectrum (carotenoids) to those absorbing in the green (phycoerythrin); thence to those absorbing in the orange (phycocyanin); and ultimately to those absorbing in the red (chlorophyll *a*) (Fig. 12.4). In each transfer, some electronic energy is converted into vibrational energy (and then dissipated into thermal energy).

We shall now consider experimental evidence of this kind of excitation transfer in vivo.

The first relevant observation was made in 1943 by H. J. Dutton, W. M. Manning, and B. M. Duggar at the University of Wisconsin. They measured the yield of fluorescence of chlorophyll *a* in a diatom, using monochromatic excitation, and found that this yield was almost the same whether the incident light was absorbed by chlorophyll *a*, or by fucoxanthol. This suggested that most light quanta taken up by fucoxanthol were transferred, by resonance, to chlorophyll *a*. Subsequently, C. S. French and co-workers in California, and L. N. M.

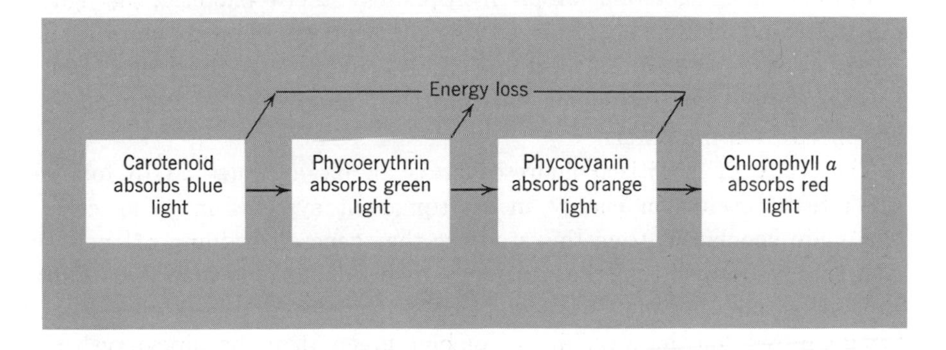

FIG. 12.4 Energy transfer from accessory pigments to chlorophyll *a*.

Duysens in Utrecht in the Netherlands, found that a similar transfer goes on, with varying effectiveness, between all accessory pigments in photosynthesizing cells and chlorophyll *a* as the final energy acceptor. Least efficient is the transfer from the yellow carotenoids; only 20–50 percent of the quanta absorbed by carotenoids find their way to chlorophyll *a*. Much more effective—of the order of 80–90%—is the transfer from phycoerythrin to chlorophyll *a* in red algae. In our laboratory, first S. Brody and then G. Tomita determined directly the time lag of emission of chlorophyll *a* fluorescence after absorption in phycoerythrin (or phycocyanin) as compared to its emission after absorption in chlorophyll *a* itself. The transfer times were found to lie between 0.3 and 0.5 nsec. In other words, they are considerably shorter than the average emission times of phycoerythrin fluorescence (7.1 nsec) or phycocyanin fluorescence (1.8 nsec). Consequently, only a small amount of phycoerythrin or phycocyanin fluorescence is emitted in competition with energy transfer from these pigments to chlorophyll *a*.

Still more efficient is the transfer from chlorophyll *b* to chlorophyll *a*—probably close to 100 percent; the same must be true for transfers from one form of Chl *a* to another in the same photosynthesis unit. ("First order" transfer may contribute to this efficiency.) Because of the smallness of energy differences, the transfer between the different chlorophylls will not be entirely in one direction. An equilibrium distribution of excitation energy will be established between the different forms of chlorophyll, such as Chl *b*, Chl *a* 670, Chl *a* 680, Chl *a* 690. Most of the fluorescence emitted from this system appears, however, to originate in Chl *a* 680 (see Chapter 15).

In the case of bacteriochlorophyll in photosynthetic bacteria, the light quanta absorbed by two forms of this pigment (those with absorption bands at 800 nm and 850 nm) are transferred to the third one, with the absorption band at about 890 nm, and practically all fluorescence is emitted from the latter.

The study of sensitized fluorescence is an elegant method to follow the fate of excitation energy in multipigment systems in plant cells. The main conclusion from this study is that general draining of excitation energy occurs from the accessory pigments to chlorophyll *a*. This transfer is, however, not fast enough to prevent some fluorescence from being emitted "on the way," in red or blue-green algae, by phycoerythrin and phycocyanin. If (see Chapter 13) the chloroplast pigments are divided into two "systems," energy migration should be analyzed separately for each of them; but this has not yet been attempted.

HOMOGENEOUS ENERGY TRANSFER

We now return to energy transfer between identical pigment molecules, particularly between chlorophyll *a* molecules in "photosynthetic units." These units may contain about three hundred chlorophyll *a* molecules. Energy transfer is known to occur in such dense assemblies; but, unfortunately, we have no simple way to demonstrate it (as sensitized fluorescence permits us to demonstrate the occurrence of heterogeneous transfer).

Since the molecules in the unit are identical, conditions in it are appropriate for a "first order" resonance interaction, with a transfer rate proportional to r^{-3}.

The consequences of this resonance interaction depend on its strength. In one extreme case—that of "weak coupling"—a proper approximation may be the picture of excitation energy migrating by a sequence of "jumps" between adjoining molecules—a kind of "random walk" of the excitation quantum. On the opposite end, that of "strong coupling," one must instead imagine a simultaneous, "communal" excitation of all pigment molecules involved in the exchange (forming, as it were, a single "supermolecule"). The latter case is well known from the study of certain dye polymers, in which strands of several hundred molecules behave like a single giant molecule. In the excited state, the electron sweeps through the whole strand. "Weak" and "strong" interaction are defined by comparison of the interaction energy with the vibrational energy in the individual molecules. If the interaction energy is smaller than the vibrational energy in the individual molecules, the absorption process occurs, in the first approximation, within an individual molecule, and the intramolecular vibrations (whose excitation converts an absorption line into an absorption band, as described in Chapter 10) are excited in the same way as in an isolated molecule. In the second case, when the interaction energy is much larger than the vibrational quantum, the excitation exchange frequency, ν_e, is (according to the basic law of quantum mechanisms, $E = h\nu$) higher than the vibration frequency, so that numerous energy transfers can take place during a single vibration. Under these conditions, electronic excitation becomes a "communal" phenomenon, and intramolecular vibrations are uncoupled from electronic excitation, with consequent far-reaching change in the shape of the absorption band.

In the case of chlorophyll *a* in vivo, the similarity of the absorption spectrum of the pigment in vivo with its spectrum in solution (in which absorption occurs in isolated molecules), clearly indicates that conditions in photosynthetic units are *not* those of "strong interaction" in which individual molecules are combined into a supermolecule. Rather, the situation approaches the case of "weak interaction," in which excitation moves around from molecule to molecule in a random walk.*

The random walk of excitation in the photosynthetic unit energy ends when the quantum is either reemitted as fluorescence by one of the molecules it visits on the way, or is dissipated by internal conversion in one of them, or reaches a "trap" (a spot where its energy is used to bring about a chemical reaction).

We believe this is what actually happens in the photosynthetic units in chloroplasts; but nobody has been able to devise an experiment directly demonstrating the random walk of the quantum in them. One indirect evidence is almost complete depolarization of chlorophyll *a* fluorescence in vivo (see below).

Simple arguments can be adduced in favor of migration of the excitation quantum as an indispensable step in photosynthesis. In order to utilize efficiently solar radiation, plant cells must strongly absorb it (compare p. 102). A leaf, containing a few layers of green cells, in fact absorbs red and blue-violet light almost completely. Even a single Chlorella cell, about 5×10^{-4} cm thick, absorbs up to 60% of incident light in the maximum of the red absorption band—this is why it appears distinctly green under the microscope. To achieve such absorption, it is not enough for the cells to contain a strongly absorbing organic pigments, such as chlorophyll (or a phycobilin); these pigments have to be present in large amounts. A French proverb says, "the most beautiful girl cannot give more than what she has"; and the most intensely colored pigment cannot absorb more quanta than its absorption coefficient permits!

* Weak first order coupling should not be confused with the still weaker second order coupling.

The highest absorption coefficients of organic pigments are of the order of 10^5 (See[3]). Typical green plant cells have linear dimensions of the order of 10^{-3} cm. To produce a 50% absorption in the band maximum ($I = 0.5I_0$), the pigment concentration in such a cell must be of the order of 3×10^{-3} mole/liter, as the following simple application of Beer's law proves. If I is equal to $0.5I_0$, we can write.

$$\log (I_0/I) = \log 2 = 0.3 = \alpha c x = 10^5 c 10^{-3}; \text{ or, } \quad c = 3 \times 10^{-3} \text{ mole/l} \tag{12.1}$$

Actually, fully green leaves do contain up to five percent of chlorophyll a in relation to dry weight, corresponding to a cellular concentration of the order of 10^{-2} mole per liter.

The purpose of this estimate was to make clear that by the very nature of the task placed before photosynthesizing cells, they must be densely packed with pigment molecules. Now, having absorbed the light quanta, the cell must use the absorbed energy to set in motion chains of enzymatic reactions, by which the unstable primary photochemical products are converted into the final products, O_2 and (CH_2O). For this, at least a dozen, if not more, specific enzymes are needed (see Chapter 17). An enzyme molecule is a protein with a molecular weight (and thus also space requirement) a hundred or thousand times larger than that of a chlorophyll molecule. There is not enough space in the cell for a separate "enzymatic assortment" to be assigned to each pigment molecule! A large number of pigment molecules simply *must* share a common enzymatic "conveyor belt."

Fortunately, the catalytic capacity of enzymes is quite sufficient for this purpose. It is easy to calculate, from Beer's law, that a single pigment molecule, with an absorption coefficient α will absorb, in a second, out of a light flux $I_{h\nu}$, (measured in number of quanta falling each second upon one cm^2), a number, n, of quanta determined by the following equation :

$$n = 4 \times 10^{-21} \alpha I_{h\nu} \tag{12.2}$$

[3] A molar absorption coefficient of 10^5 means that 6×10^{23} molecules have a total "opaque" cross section of $10^5 \times 10^2$ cm the factor 10^2 appears because concentration is measured in moles/liter). This means $10^7/6 \times 10^{23} = 1.6 \times 10^{-17}$ cm^2, or 15 A^2, per molecule—which is close to the "kinetic" cross section of a medium-sized organic molecule, such as chlorophyll, as determined, for example, from its diffusion coefficient. In other words, pigment molecules are practically "black" to light in the peak of their absorption band.

The strongest flux to which plants are exposed in nature is that of full sunlight at noon, which corresponds to $10^{16} - 10^{17}$ (visible) quanta per sec per cm^2. This means according to Eq. 12.2 that each pigment molecule will absorb, under natural conditions, only a few quanta ($n = 4 - 40$) each second. This is very little by standards of enzymatic efficiency; many enzyme molecules can handle tens of thousands of substrate molecules per second. One "enzymatic conveyor belt" is, therefore, quite sufficient to handle the photochemical production of hundreds, if not thousands, of pigment molecules.

We have already reviewed in earlier chapters (Chapters 6 and 8) evidence suggesting the presence in chloroplasts of one "reaction center" for about 300 chlorophyll molecules. A mechanism is required by which light quanta absorbed by several hundred chlorophyll molecules can be put to work on a single enzymatic conveyor belt—a system of "messengers" connecting the many pigment molecules in a unit with a single reaction center. These messengers could conceivably be material particles, for example, the primary oxidation or reduction products, diffusing from the many pigment molecule where the quantum had generated them, to a single "reception point" of the enzyme system. A more compact, safer, and faster mechanism is, however, physically possible: instead of a messenger with a letter, the communication can be by a telegram! This is what resonance energy migration amounts to. In it, only excitation energy, in the form of "excitons"—in other words, only the light electrons and not the heavy atoms or molecules, need to move.

It is the nature of a quantum phenomenon that the energy of a migrating "exciton" is not dissipated during migration, but kept in one piece. The *course* of migration can be represented by a spreading *wave,* but the whole energy of the "exciton" can be found, at a given moment, in one or the other of the many resonating molecules. The wave describes merely the spread of the *probability* of finding the quantum in different locations. (In ordinary mechanical resonance, on the other hand, the vibrational energy actually spreads over the assembly of resonating forks or bells, and is unlikely to be found assembled again in a single one of them.)

One can visualize the exciton as consisting of an excited electron and an "electron hole" left in an atom or molecule. In "random walk" of energy—the case we consider here—the electron and the hole move together, always remaining in the same atom or molecule (Fig. 12.5).

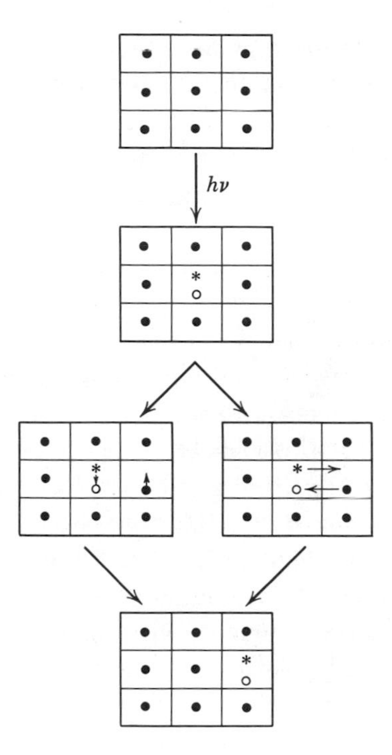

FIG. 12.5 Two mechanisms of exciton migration in a crystal lattice. *Left:* Heller-Marcus mechanism; *right:* Wannier mechanism. Dots represent unexcited molecules, asterisks, excited electrons, and circles, positive cores left after excitation.

In other words, this type of migration involves no separation of positive and negative charges.

The random walk picture of energy migration reminds one of a pinball machine, into which a steel ball had been shot; the different bulbs light up one after another, as the ball runs around on the board, until it either falls to the bottom (which corresponds to reemission of the quantum by fluorescence), or falls into a "pay hole," that is, its energy is utilized for a photochemical process.

It is worth mentioning here that resonance energy migration does not in itself affect the natural lifetime of excitation. A man could not extend his natural life expectancy by sleeping each night in a different bed! However, in practice, the random walk of the "exciton" involves risks. Some "beds" on its way may be infected, or have to be shared with dangerous companions. A migrating quantum has a greater chance of getting into such a dangerous situation than a stationary one, and this may lead to its premature death. More specifically, some molecules in the array may be engaged in chemical interactions. This can create a "trap" (or "sink") into which the migrating quantum stumbles. In other words, resonance energy migration can lead to the abbreviation of the lifetime of the exciton, and thus to the *quenching of fluorescence*. In photosynthesis, the trap may be identical with the "reaction center," where the migrating quantum is caught and put to useful work.

Resonance migration accounts, at least in part, for the well-known phenomenon of *concentration quenching* of fluorescence: strong solutions of fluorescent pigments fluoresce weaker than the more dilute ones. An even more sensitive index of resonance energy migration is *depolarization* of fluorescence—weakening or disappearance of polarization normally present in fluorescence excited by polarized light. If fluorescence is excited by plane-polarized light, that is, light that vibrates (and causes, upon absorption, the electrons in the molecule to vibrate) preferentially in a certain plane, the emitted fluorescence also is polarized. This is so because in the interval between absorbing a quantum and reemitting it, the molecule does not have enough time to forget the orientation it has had at the time of absorption. But if the energy quantum undergoes, between absorption and emission, a series of resonance transfers, each molecule in the resonance chain will be oriented somewhat differently, and after a few transfers, the original preference for a certain direction will be lost. It has been observed, for example, that the fluorescence of phycobilins, excited by polarized light, is completely depolarized. This can be considered as evidence that a quantum absorbed by one of many phycobilin molecules contained in the pigment-protein complex, moves, by resonance, from one of them to another and becomes completely "disoriented" by the time of reemission.

ELECTRON MIGRATION

A few words should be said about another possible mechanism of energy migration: the migration of *electrons*. This is a phenomenon well known from the study of crystals.

In *metallic* crystals, the electrons move freely even without external

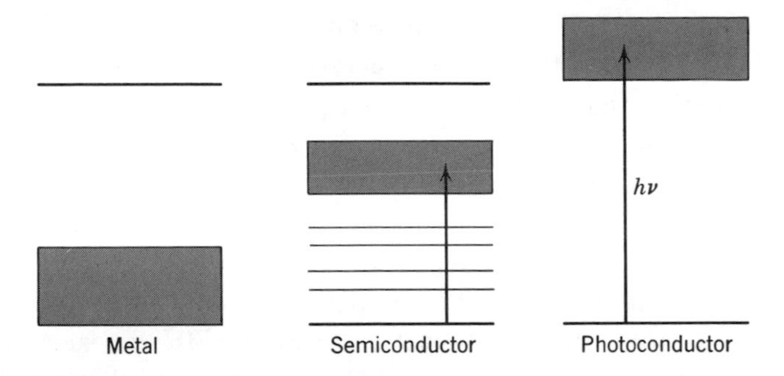

FIG. 12.6 Conductance levels (shaded areas) in a metal, a semiconductor, and a photoconductor.

excitation. The sharp energy levels of individual atoms are replaced, in metals, by so-called "conductance levels," which are not sharp, but form broad bands (Fig. 12.6). This sharing of electrons by all atoms in a piece of metal is the reason for its electrical conductivity. If we apply a potential difference to the two sides of a metallic piece, all free electrons rush in the direction of the applied force, toward the positive electrode, like kids left by their mothers to play in a park will run towards the street where a band is passing.

In *semiconductors,* such electron migration becomes possible only after thermal excitation. The conductance levels in them are higher than the energy levels in which the electrons belong to individual atoms; but the difference is small, so that moderate heating is enough to send some electrons into conductance bands. Finally, in *photoconducting* crystals, the separation between the "atomic" and the "conductance" levels is much larger than kT at room temperature (k is the Boltzman constant and T, the absolute temperature); this gap cannot be bridged by thermal excitation, but only by absorption of a light quantum. A normally insulating crystal thus becomes a conductor when exposed to light (Fig. 12.6). Such "photoconductivity" is found in many organic crystals.

In addition to conductivity based on free movement of electrons (which are negatively charged), there exists a second mechanism of photoconductivity: the migration of (positively charged) holes, left in the lattice by electrons raised into conductance levels. Migration of holes is, in essence, still a migration of *electrons:* a hole in an atom, left by excitation of an electron, is filled up by an electron supplied by a neighboring atom, and so on, in a chain. The net result is that the positive hole migrates through the crystal. If an electric potential is applied, this migration becomes directed preferentially toward the negative electrode, and the hole emerges from the crystal at this electrode. The hole migration may become more significant than the migration of the excited electrons if the latter get stuck—as electrons often do—in some traps in the lattice (provided by lattice irregularities).

The arrangement of chlorophyll molecules in the chloroplasts is *not* crystalline, as proven by its absorption spectrum. It was suggested that light absorption could nevertheless make the chlorophyll layer "photoconducting." The migration of an electron (or of a corresponding hole) could serve then as a "quantum gathering" mechanism, instead of a migration of excitons. An "enzymatic center" could represent a "sink" or "trap" catching the migrating electron (or hole).

We recall that taking up an electron is *reduction,* and losing one, *oxidation.* An electron "trap," surrounded by closely packed pigment molecules, could catch electrons produced by light absorption anywhere in the pigment body, while holes, migrating in their turn, could be trapped somewhere else, in "reducing" (that is, electron-donating) traps. This view is advocated particularly by William Arnold at Oak Ridge, who described a number of experiments suggesting photoconductive behavior of chloroplasts.

A combination of the two mechanisms—exciton migration and electron migration—has also been suggested: An exciton, produced by the absorption of a quantum, migrates through the photosynthetic unit, until it meets an oxidizing (electron-catching) or a reducing (electron-donating) enzymatic center. There, the excited electron is trapped (or the hole is filled up), and an enzymatic conveyor belt is set into motion. The exciton is now reduced to a free electron (or a free hole), which, in turn, migrates through the units until it reaches an electron-donating (or electron-catching) center. The primary photochemical oxidation-reduction process is thus completed, the oxidation product being formed in one enzymatic center, and the reduction product in another center.

This indirect mechanism of charge separation is more plausible than a direct one for the following reason: Whenever charge separation and photoconductance are a direct result of the absorption of a quantum, the absorption spectrum of the material shows that sharp excited electronic levels, characteristic of isolated molecules, have been replaced by broad conductance bands. No evidence of such transformation appears in the spectrum of chlorophyll in vivo; the absorption bands of chlorophyll in the living cell are similar to those of chlorophyll in a molecular solution. This objection does not apply to the above-described indirect mechanism of charge separation.

It is useful to realize that the difference between the "pure" exciton migration mechanism, and a combined exciton-electron (or exciton-hole) migration mechanism consists in spatial separation of the loci of primary reduction and primary oxidation in the second mechanism; while in the first one, the two processes must follow each other in the same place.

Much further experimental and theoretical work remains to be done before the role (and the mechanisms) of energy migration and electron migration in photosynthesis will be completely understood.

Chapter 13

The Two Photochemical Systems; The Red Drop and the Emerson Effect

THE RED DROP; "ACTIVE" AND "INACTIVE" CHLOROPHYLL *a*

In discussing the action spectra of photosynthesis in Chapter 11, we postponed dealing with the decline of the quantum yield (the "red drop"), which occurs above 680 nm in green algae and above 650 nm in the red ones (Fig. 13.1). (A similar drop was found, in our laboratory, also in the action spectrum of the Hill reaction.) Why should the quantum yield of photosynthesis decline in the far red, although quanta are still absorbed there by chlorophyll *a*? The first naive suggestion was that in the far red the quanta simply become "too small" to sensitize photosynthesis. This smallness is, however, irrelevant to our problem, as shown by Fig. 13.2.

Absorption at the red end of the band originates in strongly vibrating molecules in the ground state (arrow A' in Fig. 13.2), while absorption in the peak of the band (arrow A'') originates in the nonvibrating ground state. The former is weak compared to the latter because there are few strongly vibrating molecules at room temperature; but once absorption has occurred, there is no reason why the excited molecules should

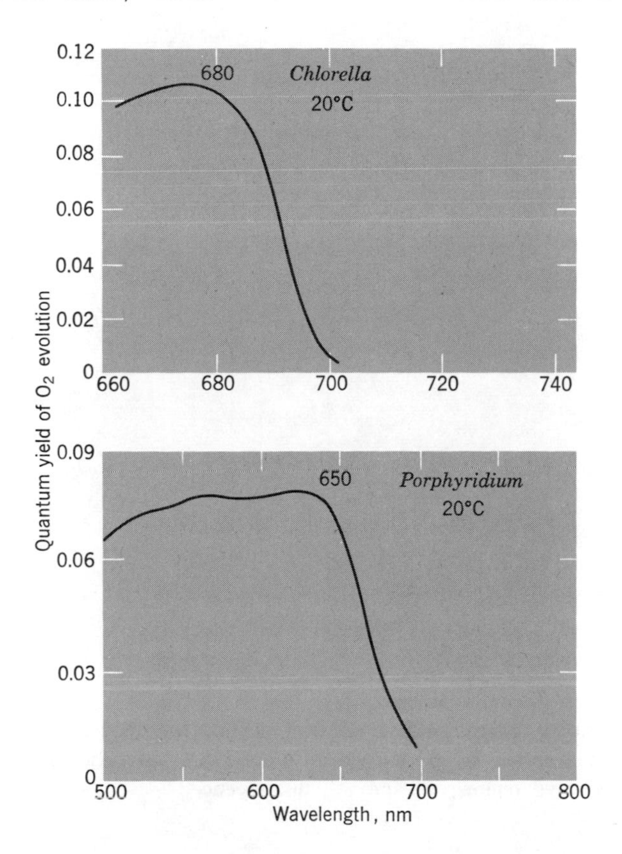

FIG. 13.1 "Red drop" in action spectra of photosynthesis [$\Phi O_2 = f(\lambda)$] in green and in red algae. (R. Emerson, R. V. Chalmers, and C. Cederstrand, 1957, and M. Brody and R. Emerson, 1959, respectively.)

behave differently from those originated in absorption from weakly vibrating states, for the following reason.

It was noted in Chapter 10 that a molecule lives long enough in the excited electronic state to be exposed (particularly in a condensed medium) to numerous encounters with other molecules. These encounters lead to energy exchange, and quickly (that is, within a period much shorter than the electronic excitation lifetime of more than 10^{-9} sec) equilibrate its "vibrational temperature" with that of the medium (wavy arrows in Fig. 13.2). Thus, by the time the excited molecule participates in a photochemical reaction, it has forgotten, as it were, the vibrational

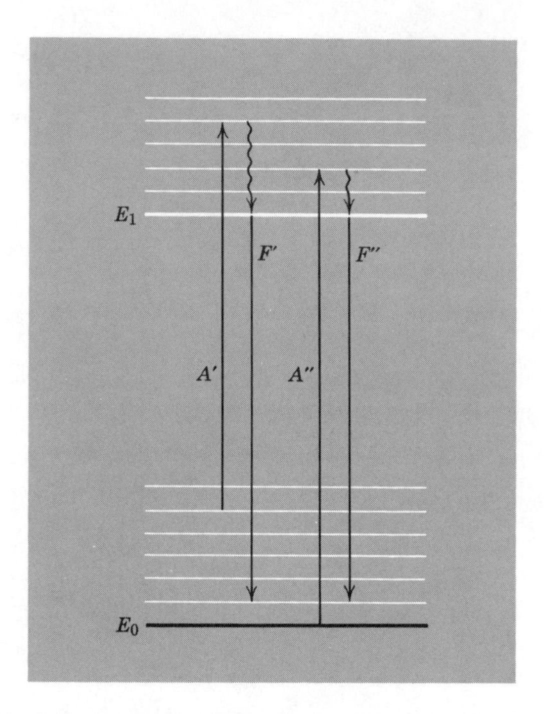

FIG. 13.2 Yield of fluorescence should not depend on frequency of light absorption. (E_0, ground state; E_1, excited state; A' and A'', absorption in the long wave and the short wave region; F' and F'', fluorescence.)

state in which it had been immediately after the excitation and thus also its origin in absorption in a vibrating (or a nonvibrating) ground state.

For this reason, the quantum yield of photosynthesis (as well as that of fluorescence) should be the same throughout an absorption band (that is, there should be no "red drop"), if this band resulted from a single electronic transition in a single type of molecules, differing only by their vibrational state.

Thus, the problem arises: Do all parts of the red band of chlorophyll *a* in vivo really correspond to one and the same electronic transition in one and the same molecule?

Several hypotheses have been proposed as to why this may *not* be true. One suggestion was that the main chlorophyll absorption band, due to single (monomeric) pigment molecules, Chl, may overlap at its long-wave end with a band due to double (dimeric) molecules, Chl_2, and that the "red drop" occurs when absorption is due to the dimers. This assumption provides a satisfactory explanation of changes that absorption and fluorescence of many dyes undergo in solution with increasing concentration (which favors dimerization). Associated dye molecules usually are nonfluorescent; the coupling of molecules in a dimer seems to permit a more rapid internal conversion of electronic excitation into ground state vibrations. This rapid deactivation may also explain declining photochemical efficiency.

Dimerization has been observed in chlorophyll solutions, although only at relatively high concentrations (about 10^{-3} molar or more). It has been suggested, particularly by A. A. Krasnovsky in Moscow and by S. S. Brody (in our laboratory, and later in New York), that dimerization provides an adequate explanation for the "red drop" in the yields of photosynthesis and fluorescence of chlorophyll in living cells; but this explanation encounters considerable quantitative difficulties.

Alternative (or additional) explanations of the complexity of the red absorption band of chlorophyll *a* in vivo have been suggested. Chlorophyll *a* may be present, in vivo, in slightly different chemical forms, with slightly displaced absorption bands and with different photochemical capacities and fluorescence yield. The difference may be in the pigment molecule itself, or in its association with different partners, such as proteins or lipids. The red drop may result from the presence, on the long-wave end of the main absorption band, of an absorption band (or bands) due to a chemically different form (or forms) of chlorophyll *a*, or to different chlorophyll complexes (rather than to chlorophyll polymers, as in the preceding hypothesis).

If one ascribes the red drop to the overlapping of bands belonging to two forms of chlorophyll *a*, a short-wave form that is "active" in photosynthesis, and a long-wave "inactive" form, one can derive, from the course of the red drop curve, the shape of the absorption bands of the two components. One arrives, in this way, at the conclusion that—at least in green algae—the "inactive" form is much less abundant than the "active" one. Its absorption peak must lie at about 695 nm (while that of the "active" form is located at about 675 nm). In red algae, on the other hand, most of the chlorophyll *a* seems to belong to the "inactive" form.

These conclusions bring to mind the analysis of the absorption band of Chl *a* in vivo into Gaussian components, described in Chapter 9. It suggested the presence, in the red band, of two main components, of about equal intensity, Chl *a* 670 and Chl *a* 680. Some evidence was mentioned also for the existence of a third, relatively weak component, Chl 695 nm (see Fig. 9.7). In green plants, this third component is likely to be the one responsible for the red drop, while the two main components contribute efficiently to photosynthesis and to fluorescence. Again, the situation must be different in red algae, where the larger part of Chl *a* 670 and Chl *a* 680 appears to be "inactive."

It will be shown later that additional facts, indicating the distribution of photosynthetic pigments between two "pigment systems," which are both active, but play a complementary role in photosynthesis, call for a more sophisticated interpretation of the red drop.

THE EMERSON EFFECT; THE HYPOTHESIS OF TWO PHOTOCHEMICAL REACTIONS AND TWO PIGMENT SYSTEMS

In making a closer study of the "red drop" in the quantum yield of photosynthesis in Chlorella, Emerson and co-workers found, in 1957, that the "inefficient" far-red light (i.e., in green cells, light beyond 680 nm) can be made fully efficient by simultaneous illumination with light of a shorter wavelength—for example, by combining a beam of 700 nm light with a beam of 650 nm light (Fig. 13.3). Oxygen production from the two combined beams was found to be substantially higher than the average of the sum of the production from the separate beams; and since it was the far-red beam that, used by itself, gave a "subnormal" yield, it was natural to interpret the result as evidence of improvement of the photosynthetic efficiency of far-red light (700 nm) by simultaneous illumination with orange-red light (650 nm). This enhancement became known as the Emerson effect. It can be expressed as the ratio of the rate of oxygen evolution (ΔO_2) in the far-red light in the presence of the supplementary beam, and the same rate in the absence of it:

$$E = \frac{\Delta O_2 \text{ (in combined beams)} - \Delta O_2 \text{ (short-wave beam alone)}}{\Delta O_2 \text{ (long-wave beam alone)}} \quad (13.1)$$

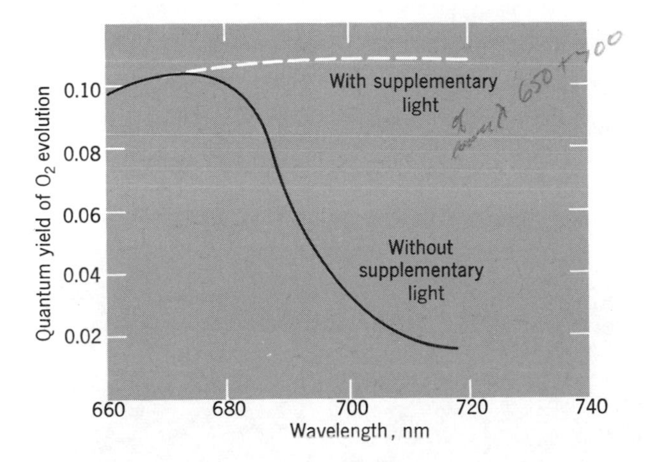

FIG. 13.3 The Emerson effect in *Chlorella*.

Emerson and co-workers studied the action spectrum of this effect; i.e., they made plots of E versus the wavelength of supplementary light (Fig. 13.4). Enhancement proved to be greatest when the largest proportion of supplementary light was absorbed by one of the accessory pigments—chlorophyll b in green algae, chlorophyll c or fucoxanthol in diatoms, and phycobilins in red and blue-green algae. It thus seemed that what was needed to make photosynthesis fully efficient, was the absorption, in addition to a quantum taken up by chlorophyll a, of another quantum taken up by one of the accessory pigments! The red drop takes place in the spectral region where chlorophyll a takes over as the only absorber.

The interpretation of the early drop in red algae appeared particularly easy in this picture. These algae contain no chlorophyll b, and the region of exclusive absorption by chlorophyll a extends in them down to about 650 nm, where the absorption by phycocyanin becomes significant; and this is where the red drop begins!

These observations caused Emerson to suggest tentatively that photosynthesis involves two photochemical reactions—one sensitized by chlorophyll a and one sensitized by accessory pigments.

The occurrence of two light reactions is now generally accepted; but the assumption of direct sensitization of one of them by accessory pigments runs into difficulties (apart from a certain implausibility of assign-

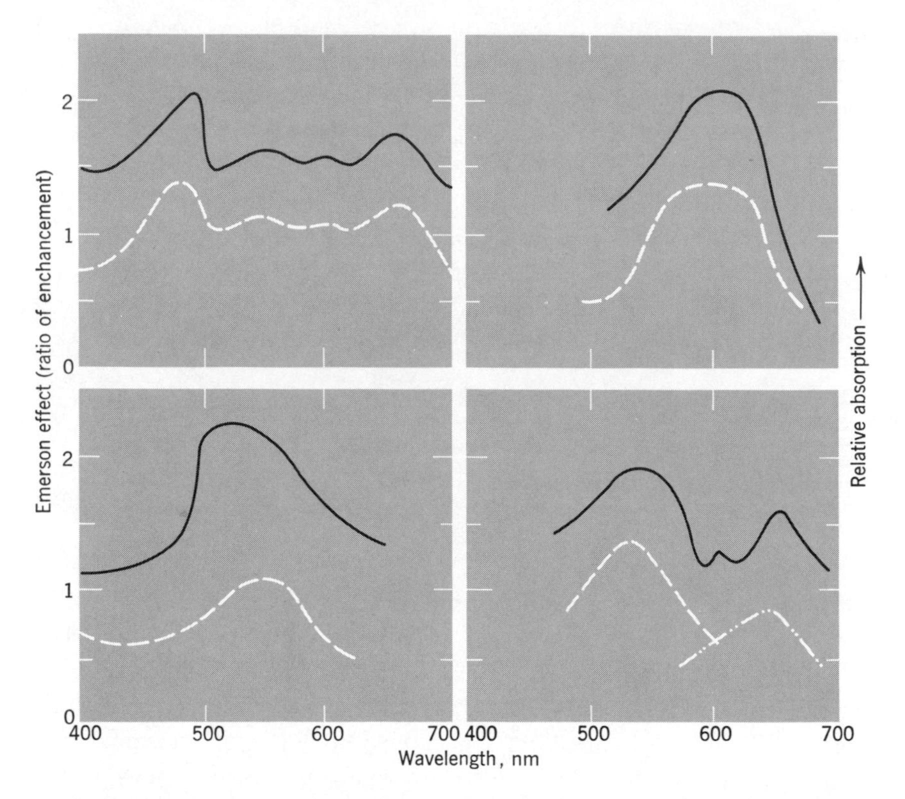

FIG. 13.4 Action spectra of the Emerson effect in different algae. The green alga *Chlorella pyrenoidosa* (top left): the blue-green alga *Anacystis nidulans* (top right), the red alga *Porphyridium cruentum* (bottom left), and the diatom *Navicula minima* (bottom right). In each case, the solid curve shows the action spectrum of the Emerson effect, *i.e.*, the degree enhancement as the wavelength of the supplementary illumination is varied. These curves turn out to be approximately parallel to the absorption curves (broken lines) of the various accessory pigments: chlorophyll *b* in *Chlorella,* phycocyanin in *Anacystis,* phycoerythrin in *Porphyridium,* and fucoxanthol (broken) and chlorophyll *c* (dotted broken) in *Navicula.* (R. Emerson and E. Rabinowitch, 1960.)

ing a common photochemical function to all the, chemically very different, accessory pigments). How can Emerson's suggestion be reconciled with observations of phycobilin-sensitized fluorescence of chlorophyll *a* in red algae (see Chapter 12)? Occurence of sensitized fluorescence clearly shows that energy quanta absorbed by the accessory pigments are transferred, often very efficiently, to chlorophyll *a*, causing the latter to emit fluorescence. How could quanta absorbed by an accessory pigment have a different photochemical function from quanta absorbed by chlorophyll *a*, if they are transferred quantitatively to the latter? This makes no sense!

In 1960, we observed in the action spectrum of the Emerson effect, in addition to peaks attributable to accessory pigments, also a peak at 670 nm, that is, within the main absorption band of chlorophyll *a* itself (Fig. 13.5). This changed the picture: what now appeared to be necessary for effective photosynthesis, was absorption of one quantum in one form of chlorophyll *a* and of another quantum in another form of the same pigment! The peaks in the action spectrum of the Emerson effect, corresponding to accessory pigments, could be reinterpreted as due to preferential resonance energy transfer of excitation energy from the accessory pigments to the "active" form of chlorophyll *a*.

This generalization appeared more plausible than Emerson's original

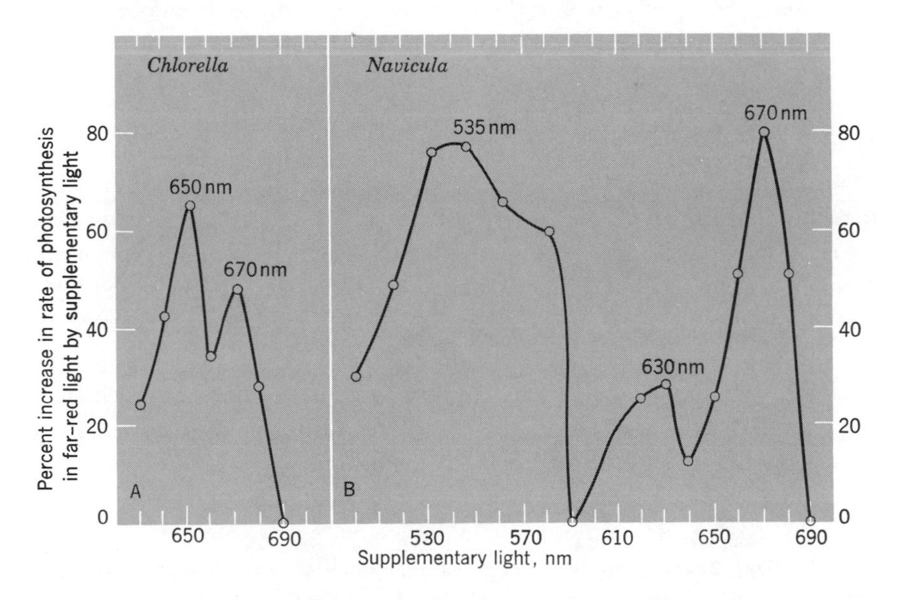

FIG. 13.5 Details of action spectra of the Emerson effect in a green alga (*Chlorella*) and in a diatom (*Navicula*), showing a peak at 670 nm; the peak at 650 nm is due to Chl *b*, at 535 nm to fucoxanthol, and at 630 nm, to chlorophyll *c*. (Govindjee and E. Rabinowitch, 1960.)

one. It required, however, giving up the above-suggested simple explanation of the early red drop in red algae. In the spectrum of the Emerson effect in these algae, only peaks due to phycobilins were observed, and no peak at 670 nm; and yet there is no evidence of a deficiency of Chl *a* 670 in these organisms.

A provisional picture, taking into account all results, can be attempted along the following lines:

All photosynthetic pigments are distributed in two "pigment systems" (Fig. 13.6), which, following a suggestion by Duysens, we will call "system I" (PSI) and "system II" (PSII). System II is (relatively) strongly fluorescent; system I only very weakly fluorescent. System II contains, in green plants, both Chl *a* 670 and Chl *a* 680, with a slightly larger proportion of the former, and no Chl *a* 695. System I also contains both Chl *a* 670 and Chl *a* 680, but somewhat more of the latter; most importantly, it contains Chl *a* 695 (perhaps, 10% of total chlorophyll *a* in this system). The latter drains, by resonance transfer, practically all quanta from Chl *a* 670 and Chl *a* 680 in this system and effectively dissipates them—probably by internal conversion—allowing only a very low yield of fluorescence in this system.

The assumption that there is relatively more Chl *a* 670 in system II than in system I is needed to explain the appearance of a peak at 670 nm (Fig. 13.5) in the action spectrum of the Emerson effect in Chlorella and in Navicula. In red algae, we must assume that PSII consists in a larger proportion of phycobilins, with only a very small amount of Chl *a*, while PSI contains the bulk of Chl *a* (including Chl *a* 670 and Chl *a* 680, and also Chl *a* 695) and a smaller proportion of the phycobilins.

If the two pigment systems sensitize two consecutive photochemical steps in photosynthesis, as it is now widely postulated, they must operate at equal speed. At each wavelength, some energy is absorbed in one and some in the other pigment systems. The amounts, however, are unequal. Above 650 nm (in red algae) and above 680 nm (in green algae), "too much" energy seems to go to system I, and too little to system II, causing the "red drop" in the yield of photosynthesis. The Emerson effect reveals improvement of this distribution by simultaneous illumination with a second light beam. The fact that photosynthesis does occur, albeit with low efficiency, throughout the region of the red drop

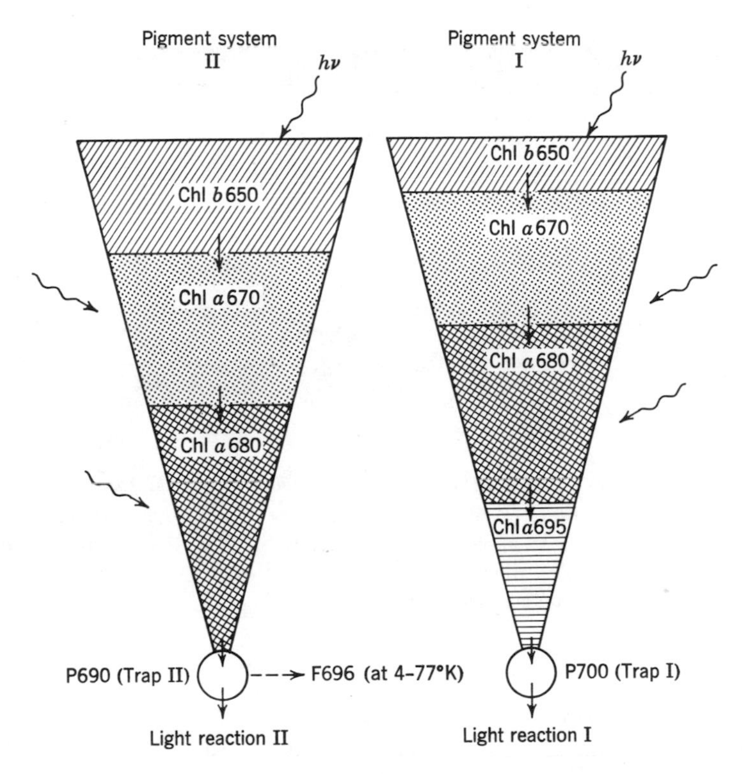

FIG. 13.6 A working hypothesis for the distribution of the chlorophylls in the two pigment systems in the higher plants and the green algae. The two systems seem to contain both chlorophyll *a* (Chl *a*) and chlorophyll *b* (Chl *b* 650), but in different proportions. (In the red and the blue-green algae, the phycobilins replace Chl *b*.) It is suggested that the long-wave form of chlorophyll *a* (Chl *a* 695) is present only in the pigment system I. The two "bulk" chlorophylls *a* (Chl *a* 670 and Chl *a* 680) are almost equally distributed in the two systems, but there is relatively more Chl *a* 670 in system II. (In the red and the blue-green algae, a much larger proportion of Chl *a* is in pigment system I.) The energy trap (Trap I) of the system I is P700 and the Trap II is P690.[1] (Govindjee, G. Papageorgiou, and E. Rabinowitch, 1967).

(up to 720 nm), shows that both photochemical reactions can take place there, although one proceeds so slowly as to become a bottleneck for the whole process. At the shorter waves, both photochemical systems are excited at a much better balanced rate.

Duysens pointed out that if the two light reactions, sensitized by

[1] Witt and co-workers located this band at 690 nm in chlorella and at 682 nm in "system II particles" (compare p. 192).

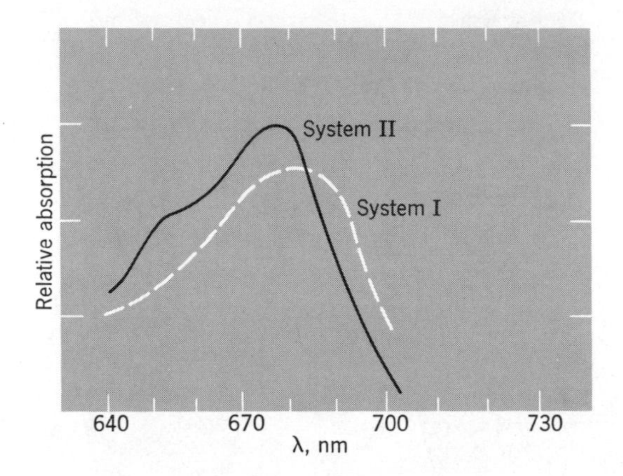

FIG. 13.7 Absorption spectra of two pigment systems derived from the data of C. S. French and co-workers (1960), on action spectra of photosynthesis in *Chlorella* cells "saturated" with light absorbed preferentially in PSI (far-red light) and in PSII (near-red light). The difference may be explained by the presence of more Chl *b* in PSII and of more Chl *a* 695 in PSI.

the two pigment systems, occur in series, one way to determine the absorption spectrum of each system is by measuring the action spectra of photosynthesis in the presence of strong (but not saturating) light belonging to the other system. The rate of the overall reaction is then determined by that of the limiting light reaction. When excess light goes to system I, the rate (and thus the action spectrum) is determined by system II, and vice versa. Such measurements had been made by French and co-workers in California; the result is shown in Fig. 13.7. This figure confirms that both pigment systems contain several forms of chlorophyll; system II seems to be much richer in Chl *b*, and somewhat richer in Chl *a* 670; while system I contains relatively more Chl *a* 680, and all of Chl *a* 695.

This is one observation—we will quote others later—suggesting that a simple identification of Chl *a* 670 as sensitizer of one photochemical reaction in photosynthesis, and Chl *a* 680 as sensitizer of the other (a tempting and simple suggestion, which we ourselves had made at first), cannot be true.

That the two light reactions are linked together by a relatively slow dark reaction first became clear from experiments in which the Emerson effect was observed to occur even when the two light beams entered the reaction vessel at right angles to each other, thus striking a given algal cell at different times (with an interval depending on the intensity of stirring). J. Myers and C. S. French noted that the Emerson effect could be observed also using alternate flashes of two beams, with intervals up to several seconds between them. In the red alga *Porphyridium*, oxygen evolution by a green flash (absorbed by phycoerythrin, belonging mainly to system II) was enhanced when it was preceded by a red flash, absorbed by chlorophyll *a* in system I, suggesting that the latter left a long-lived intermediate. One half of this intermediate appeared to survive for as long as 18 seconds (see Fig. 13.8). On the other hand, oxygen evolution from a red flash was not enhanced when it was preceded by a green flash, suggesting that green light absorbed in PSII produced no long-living intermediates.

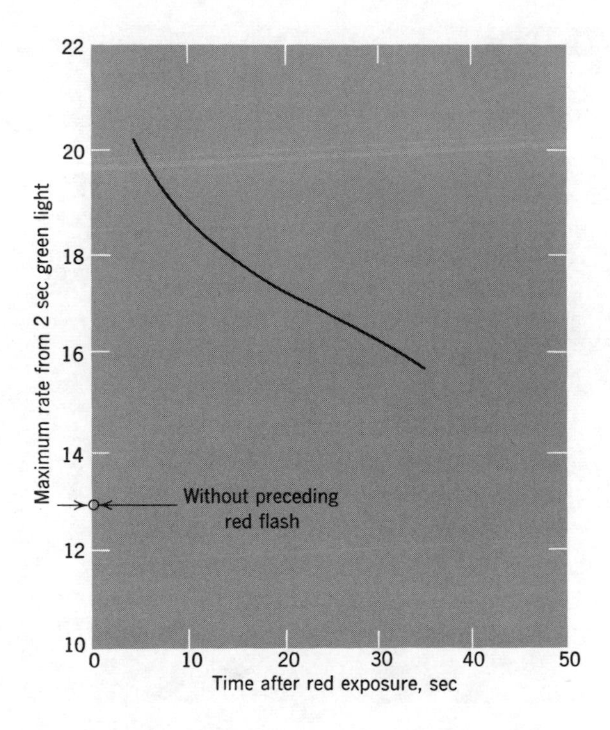

FIG. 13.8 Enhancement of oxygen yield (in red algae) from a green flash by preliminary exposure to a red flash, as function of time between the two flashes. (C. S. French, 1963.)

We are thus justified in postulating that the two photochemical primary processes in photosynthesis occur in series, with a dark reaction involving the products of the two photochemical reactions linking them together.

We shall discuss, in Chapters 14–16, other evidence that supported the conclusion that photosynthetic pigments are organized into two systems with different photochemical functions. Here, we summarize the mechanism of photosynthesis as derived from red drop and enhancement studies (Fig. 13.9).

If we assume that two separate photosystems sensitize two consecutive light reactions in photosynthesis, we must ask: How are the absorbed light quanta distributed equally between the two pigment systems permitting them to run at equal speed—as this is needed for photosynthesis

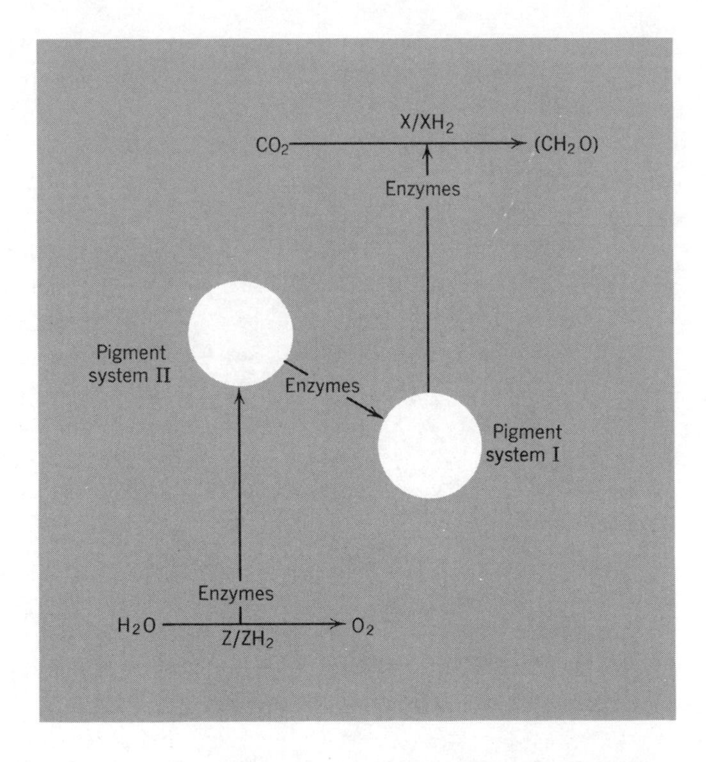

FIG. 13.9 Scheme of photosynthesis with two photochemical steps (compare with Fig. 5.4) sensitized by two pigment systems (I and II).

to proceed with the highest quantum yield? Two different hypotheses have been suggested. In one of them, called the "spill-over" hypothesis by J. Myers (University of Texas in Austin), all excess light energy absorbed in pigment system II is assumed to "spill over" into system I—but not vice versa, because quanta absorbed in PSI are drained into the long-wave component Chl *a* 695, from which their return to Chl *a* 680 and Chl *a* 670 is obstructed by an energy barrier. The spill-over leads to balanced excitation of the two systems throughout the part of the spectrum where more light is absorbed by system II than by system I, and explains the drop-off of the yield in the far-red (beyond 680 nm) where light is absorbed (in green cells) mainly by Chl *a* 695 in system I. In phycobilin-containing algae, where most Chl *a* is in system I, the drop begins at 650 nm, where absorption in phycocyanin yields to absorption by chlorophyll *a*.

This hypothesis explains the high quantum yield of photosynthesis in green cells at wavelengths $<$680 nm, and the "red drop" in the action spectrum of photosynthesis beyond 680 nm (in higher plants and green algae) and beyond 650 nm in red algae.

In the second hypothesis, called by J. Myers *"separate package"* hypothesis, an approximately balanced excitation of both systems is achieved, without any "intersystem transfer," through the presence in both systems of the same pigments, even if in somewhat different proportions. According to this hypothesis, the action spectrum of photosynthesis must show minima wherever one of the two systems absorbs more than the other. This must cause a fine structure in the action spectrum. Some such structure is present in the action spectrum of photosynthesis as well as that of fluorescence; for example, the dip in the action spectrum of photosynthesis in *Chlorella* at about 660 nm, first noted by Emerson, was clearly demonstrated as real by recent careful measurements of Rajni Govindjee in our laboratory.

Altogether, the "separate package" theory, first proposed by L. N. M. Duysens, seems at present to be more plausible than the spill-over hypothesis. The spatial separation of PSI from PSII, indicated by fractionation experiments (to be described in Chapter 16) can well explain the absence (or at least, relative weakness) of "intersystem" transfer. Furthermore, one has to take into account that the potential barrier opposing the return of energy quantum from Chl *a* 695 to Chl *a* 680 (and thus also from PSI to PSII) should not be prohibitive at 300°K.

THE HILL REACTION AND THE BACTERIAL PHOTOSYNTHESIS: TWO SYSTEMS OR ONE?

In the Hill reaction (as observed with $NADP^+$ as oxidant), the quantum requirement of 8 (per O_2 evolved), together with the occurrence of red drop and of an Emerson effect (R. Govindjee and also M. Avron, and coworkers) suggests the operation of both photochemical systems. (However, with ferricyanide as oxidant, at low light intensities, no Emerson effect could be detected in the Hill reaction.) Hill oxidants such as quinone and ferricyanide, with E_0-values of 0.0 volt (or higher) could conceivably enter the reaction sequence of Fig. 13.9 in the middle (instead of on top), so that PSII alone would be put to work, while PSI would idle. This would not decrease the quantum requirement, since one half of the light energy absorbed in the unoperative pigment system I would go to waste; and a "red drop" would still be present.

It seems that whenever the oxidation-reduction potential of the Hill oxidant is more negative than 0.0 volt, both pigment systems are needed; this is clearly the case in the Hill reaction with $NADP^+$ as oxidant. But if E_0 of the Hill oxidant is positive, one pigment system may suffice, at least under certain conditions (such as low light intensity, and sufficiently high concentration of the oxidant).

The energy requirement of bacterial photosynthesis could be easily satisfied by one light reaction; nevertheless, the quantum requirement is of the order of 8 quanta per reduced molecule of CO_2. No decline in the quantum yield at the far red end of the absorption band and no Emerson enhancement effect have been observed in bacteria. However, this does not preclude the existence of two photosystems; their absorption spectra may be simply more closely identical in bacteria than in the higher plants. Recently, Sigehiro Morita in Japan, M. A. Cussanovich, R. G. Bartsch, and Martin Kamen in La Jolla, California, and Christiaan Sybesma in our laboratory, all found evidence of the involvement of two photosystems in the photosynthesis of bacteria.

Chapter 14

Difference Spectroscopy: The Roles of the Cytochromes, P700, Plastoquinone and Plastocyanin

DIFFERENCE SPECTROSCOPY

Wc discussed, in the preceding chapter, evidence for a two-step mechanism of photosynthesis derived from the study of the action spectra of photosynthesis, in monochromatic light and in combinations of two monochromatic beams. We will now consider relevant evidence derived from experiments of another type—*difference spectroscopy*.

What one observes in this type of study is differences between the absorption spectra of plant cells in darkness (that is, in absence of photosynthesis) and in light, when photosynthesis is going on. The sample is illuminated by a strong "actinic" beam that causes the photochemical change; the change in absorption is monitored by a weak "measuring" beam. Of course, the measuring light itself, however weak, does have a certain influence on the photosynthetic apparatus; but the presumption is that this influence is so slight that the spectrum obtained in the absence of the stronger "actinic" illumination, is characteristic of nonphotosynthesizing cells. The difference spectrum, obtained in this way, reveals changes in the nature (or amount) of certain light-absorbing constituents in illuminated cells. One is particularly interested in changes that are reversed immediately upon return to darkness, since these are likely to originate in catalytic and photocatalytic components undergoing reversible changes during photosynthesis.

Several experimental difficulties present themselves in these studies. One is avoiding contamination of measuring light with actinic light. To minimize it, the two beams are sent into a rectangular vessel under right angles to each other (Fig. 14.1); but plant cells scatter light so strongly that some actinic light usually gets into the measuring beam. Complementary colored filters are often used to eliminate this contamination. For example, if the measuring light is blue, red actinic light can be used, and a blue filter inserted between the sample and the recording instrument. However, to plot the complete difference spectrum one must be able to vary the wavelength of the measuring beam through the whole visible spectrum; and in trying to do so, one gets into spectral regions where the wavelengths of the actinic and of the measuring light are so close that their separation by colored filters is impossible.

Two experimental tricks help. One is to send in the actinic light in the form of a strong flash, and to measure absorption changes after the flash. (The use of powerful condenser discharge flashes for this type of research was first introduced by G. Porter and R. G. W. Norrish, who received the 1967 Nobel Prize in chemistry for studies of extremely fast chemical reactions.) This procedure has the advantage of permitting to vary the time interval between illumination and measurement, and thus to obtain evidence as to the kinetics of the various absorption changes: how quickly they arise in light, and how rapidly (and following what kinetic law) they disappear in the dark.

Another trick is to *modulate* the measuring beam, for example, by means of a rotating disc, and arrange the measuring circuit (usually containing a photomultiplier, that is, a multistage photoelectric cell) in such a way that it will respond to modulated measuring light, but not to constant scattered actinic light (or the constant fluorescence excited by it).

Measurements of the difference spectra of photosynthesizing cells in constant light were first carried out by L. N. M. Duysens, and described in his doctoral dissertation (University of Utrecht, Holland, 1952). The flashing light technique was first developed by B. Kok, also in Holland

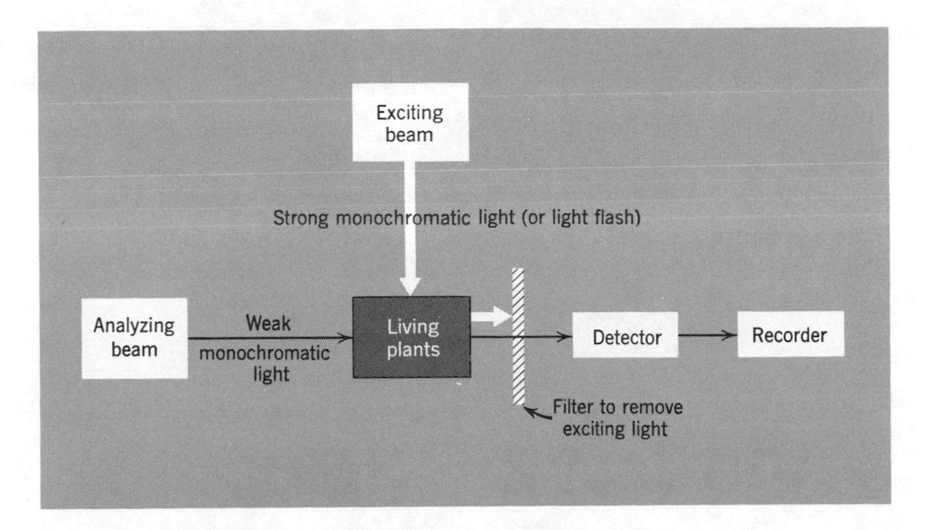

FIG. 14.1 Diagram of a difference spectrophotometer. The analyzing beam (of low intensity) is split into two beams; one goes through the sample (plant cells or living plant) and the other around the sample (not shown in the diagram). Both beams fall on the detector (a photomultiplier). The intensity of the beam that is sent around the sample is adjusted to match the intensity of the light transmitted by the sample; a zero reading is obtained on the recorder. Upon turning on the exciting beam (of high intensity), changes in absorption are observed and recorded. By varying the wavelength of the analyzing beam, a difference spectrum (that is, the difference between the absorption spectrum of cells illuminated with strong exciting light and non-illuminating cells) is obtained; it gives information concerning the pigments that undergo reversible changes during photosynthesis. By varying the wavelength of the exciting beam, an action spectrum (that is, the curve showing the effectiveness of different wavelengths of light in producing the change) is obtained; it provides information concerning the pigments that sensitize the absorption changes.

(now at Baltimore, Md.) using rotating disc flashes, and by H. T. Witt in Germany using condenser discharge flashes.

A large number of difference spectra have been obtained by the two methods, using different algae as well as chloroplast suspensions, under a variety of conditions. Figure 14.2 shows some examples.

These difference spectra reveal reversible changes in the oxidation-reduction state of several known organic compounds—three cytochromes, plastoquinone, plastocyanin, "pigment 700" and "pigment 690" (prob-

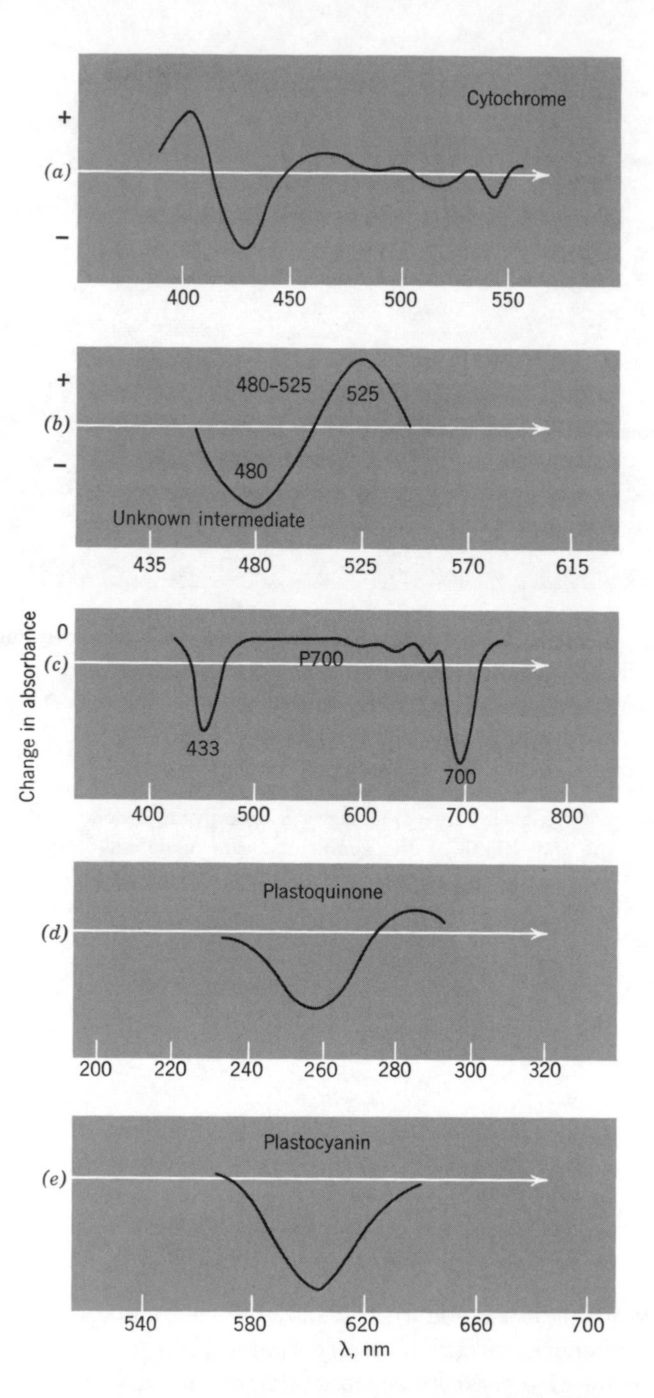

FIG. 14.2 Diagram of typical difference bands of photosynthesizing cells. Derived from the work of L. N. M. Duysens (cytochrome), B. Kok (P700), H. T. Witt, and J. Amesz (plastoquinone), and D. C. Fork and co-workers plastocyanin).

ably special forms of chlorophyll *a*)—and of some unknown ones. All these must be intermediates (or, at least, reversibly formed side products) in the reaction sequence of photosynthesis.

THE CYTOCHROMES

One hoped at first that difference spectroscopy will reveal changes in chlorophyll (or other pigments)—compounds most closely associated with the primary process in photosynthesis. Instead, Duysens' first clear-cut results (Fig. 14.2*a*) revealed, quite unexpectedly, a reversible tranformation in light of compounds whose role in photosynthesis had not been realized before—the so-called *cytochromes.* These are proteins with an attached iron-bearing prophyrin group (somewhat similar to hemoglobin). The iron in them can exist in "ferric" or "ferrous" form, and they are known to mediate, by alternating between these two forms, important oxidation-reduction steps in the respiratory chain. The cytochromes have absorption bands in the 350–600 nm range, which are generally used for their identification. These bands are different in the oxidized and the reduced state.

The two main types of cytochromes taking part in respiration are designated *b* and *c*; the former have redox potentials close to 0 volt and the latter, above +0.3 volt. When an electron is transferred from a *b*-type cytochrome to a *c*-type cytochrome, a free energy amount of about 7 Kcal/mole is liberated; two such transfers are utilized, in respiration, to produce one molecule of "high energy phosphate," ATP (Chapter 18).

In 1951, the British biochemist Robert Hill and his co-workers in Cambridge found that chloroplasts contain (at least) two cytochromes. One of them, which they called cytochrome b_6, was of "*b*-type," with a redox potential close to 0.0 volt. The other, which they called cytochrome *f*, was similar to cytochromes of the *c*-type; it had a redox potential as high as +0.42 volt. Duysens noted reversible oxidation of cytochromes (most clearly, that of cytochrome *f*) in illuminated photosynthetic cells. Since it was found that chloroplasts, freed from adhering traces of cell plasma (and thus also of mitochrondria) do not respire, that is, do not absorb oxygen in the dark, it became clear that cytochromes must play a role in photosynthesis. At first, attempts were made to place them on one or the other end of the primary photochemical process, either on the "oxidation end," close to the evolution of oxygen, or on the "reduction end," close to the reduction of carbon dioxide. However, the redox potentials of the cytochromes did not fit them into these positions; and when the concept of a two-stage photochemical process arose, the hypothesis naturally offered itself that the cytochromes are intermediates *between the two stages.* (The redox potential of the couple CO_2/(CHOH) is —0.4 volt; that of the couple O_2/H_2O, +0.8 volt; the midpoint between them is +0.2 volt, just where the redox potentials of the cytochromes are clustered.)

This concept received a convincing support from Duysens and co-workers' findings in 1961. Duysens' original observation, that cytochrome *f* is oxidized in light and reduced in darkness, proved incomplete; photooxidation occurred only in light absorbed in pigment system I (that is, in the red algae, on which these experiments were made, in chlorophyll *a*) while light absorbed in pigment system II (that means, in the red algae, in phycoerythrin) accelerated the return of oxidized cytochrome *f* into the reduced state (Fig. 14.3).

Robert Hill and Fay Bendall suggested, in 1960, that one of the two cytochromes found in plant cells (cytochrome b_6) could function (in its oxidized form) as primary oxidant in what we now call system II, accepting, in light, electrons originating in a water molecule, while the other (cytochrome *f*), could serve (in its reduced form) as electron donor (reductant) in what we now call system I, which transfers the electron to an organic compound such as a pyridine nucleotide. (The latter in turn uses it for enzymatic reduction of carbon dioxide; see Fig. 14.4.) Reduced cytochrome b_6 and oxidized cytochrome *f* could then react, directly or through additional intermediates, restoring the first one to its oxidized and the second one to its reduced state. About 0.4 eV of free energy would become available in this downward electron slide. By analogy with the similar steps in respiration, this energy could well be used to synthesize ATP. This is a very welcome possibility, because Melvin Calvin's studies with radiocarbon tracer have revealed a need for ATP in one (or two) steps in the enzymatic reaction sequence of photosynthesis (the "Calvin-Benson cycle," see Chapter 17). Since photosynthesis proceeds, in strong light, an order of magnitude faster than respiration, the needed ATP molecules can not be "borrowed" from cellular respira-

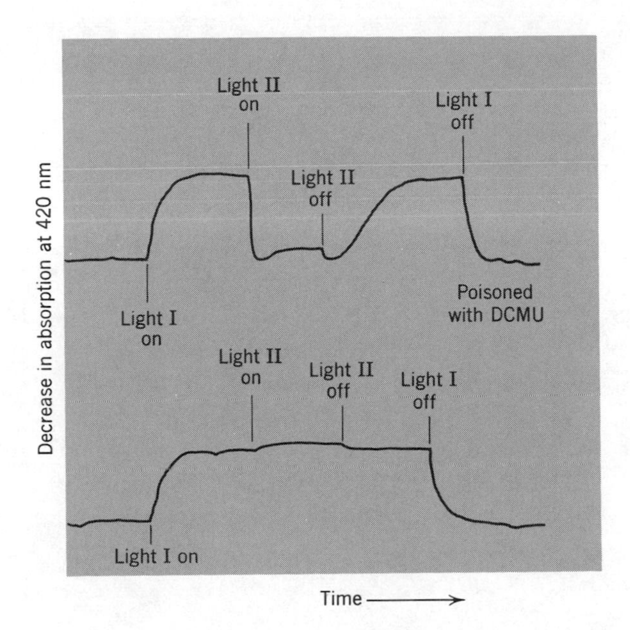

FIG. 14.3 Absorbancy changes at 420 nm owing to oxidation and reduction of cytochrome *f* in the red alga *Porphyridium cruentum*. (DCMU = 3-(3′,4′ dichlorophenyl 1,1 dimethyl urea; Light I = light absorbed mainly in pigment system I; Light II = light absorbed mainly in pigment system II.) (L. N. M. Duysens and J. Amesz, 1961.)

tion; they must be obtained as by-products of photosynthesis itself. The Hill-Bendall scheme (Fig. 14.4) provides a plausible explanation how this ATP-production can be achieved without using up additional light quanta (beyond the 8 required to move four electrons through two photochemical reactions each).

As mentioned above, Duysens' results were clearest in the case of cytochrome *f*. Hill's suggestion that a symmetric role is played by cytochrome b_6 in system II, could not be confirmed by difference spectroscopy. Also, it appeared that cytochrome b_6 may be involved in another reaction—the occasional reversal of photoreaction I accompanied by ATP-synthesis ("cyclic phosphorylation," see Chapter 18). On the other hand, recent spectroscopic evidence suggested that another *b*-type cytochrome (called cytochrome b_3) occupies the position in system II suggested by Hill and Bendall for cytochrome b_6.

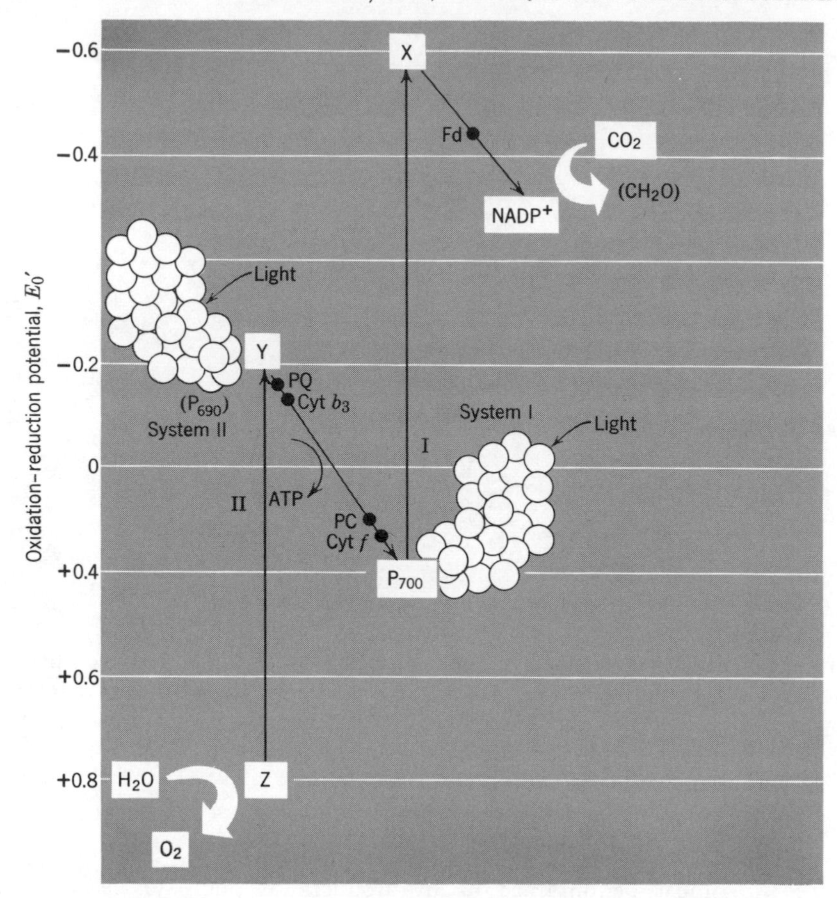

FIG. 14.4 Hill-Bendall scheme of photosynthesis in its modified form (elaboration of Fig. 13.9). Z and Y are primary electron donor and acceptor of light reaction II; P700 and X are primary electron donor and acceptor of light reaction I. P690, reaction center for light reaction II; P700, reaction center for light reaction I, PQ, plastoquinone; PC, plastocyanin; Cyt b_3, cytochrome b_3; Cyt f, cytochrome f, ADP, adenosine diphosphate; ATP, adenosine triphosphate; Fd, ferrodoxin; $NADP^+$, nicotinamide adenine dinucleotide phosphate.

THE DIFFERENCE BANDS AT 480 AND 520 nm

Reversible transformation of cytochromes in light explains only one part of the difference spectrum of illuminated plant cells. This part is prominent in weak light, but soon becomes light-saturated; while other features of the difference spectrum continue to grow with increasing intensity of illumination. Some of these changes seem to be associated with reversible changes of chlorophyll (the search for which had motivated the first application of difference spectroscopy to photosynthesis); others indicated changes in certain quinone-type compounds (see below). The origin of some of the most prominent difference bands in green plants still remains uncertain. For example, in *Chlorella,* the two most striking components of the difference spectra in strong light are a decrease of absorption (a "negative difference band") at 480 nm, and an increase in absorption (a "positive difference band") at 515–520 nm (see Fig. 14.2*b*). Quantitative studies of these spectra in Witt's laboratory in Germany, and at the University of Illinois, have dealt in particular with the 520 nm band; it is not only the most prominent, but also the best reproducible of all difference bands. This band has been taken by Witt's group as an index of chlorophyll *b* changes. Yet, no final interpretation of this band exists. Observations suggest that the difference bands at 480 and 520 nm may be composite, perhaps due to superimposed changes in chlorophyll *a,* chlorophyll *b,* and the carotenoids.

PIGMENTS P700 AND P690

Bessel Kok, using the flash method, observed (in several algae and in chloroplast fragments) an interesting difference band in the far red. It was located at about 700 nm, that is, on the long-wave side of the main absorption band of Chl *a* (Fig. 14.2*c*); Kok interpreted this band as evidence of the presence, in vivo, of a minor Chl *a* component with an absorption peak of 700 nm, (and, as he later found, a Soret band at 433 nm). This component, alone in the whole pigment system, seems to be reversibly bleached in light. He called it P700, and suggested that it represents the "trap" in the pigment system in which the light energy, taken up in a photosynthetic unit, is collected, and which alone is directly involved in the primary photochemical process. Various findings indicate that it is specifically the trap in system I, engaged in the reduction of CO_2 by electrons taken from reduced cytochrome *f*.

This suggestion is supported by the oxidation-reduction potential of

P700, which Kok determined by adding redox systems of known potential and observing their effect on the P700 difference band in chloroplast suspensions. (It is difficult to make such experiments with intact cells, because the cell membrane prevents the outside redox system from affecting the potential inside the cell.) A redox potential of $E_0' \cong +0.4$ volt was estimated for P700—about right for the position suggested for it in Fig. 14.4. We can postulate that in pigment system PSI, the pigment P700 is the primary reductant (electron donor). Following its photochemical oxidation, P700 is rapidly converted back into reduced form by a dark reaction with cytochrome *f* (or plastocyanin, see below).

Another argument in favor of identifying P700 with the "trap" is the concentration of the compound. Assuming that *all* P700 is oxidized when chloroplasts are exposed to sufficiently strong light (and also that the molar absorption coefficient of P700 is the same as that of the bulk of Chl *a*), Kok calculated that about one molecule of P700 is present per 200–300 Chl *a* molecules. (If a single "trap" were present in each photosynthetic unit in system I, a ratio of one P700 per 600 Chl *a* could be anticipated.)

Kok and co-workers found that in the blue-green alga *Anacystis,* the red light, absorbed by chlorophyll *a* in PSI, oxidizes P700; while orange light, absorbed by phycocyanin (in PSII), reduces it.

It is worth mentioning that P700 cannot be identified with Chl *a* 695. The latter was estimated to form about 5%, while P700 is present only in a concentration of 0.3% of total Chl *a*.

To sum up, we can now suggest that the photosynthetic units in PSI in green plants consist of four components: (1) some chlorophyll *b*; (2) Chl *a* 670 and Chl *a* 680; (3) Chl *a* 695; and (4) a single (or a few) molecules of P700.

Long-sought-for absorbancy changes attributable to the reaction center of pigment system II have been recently identified by G. Döring, H. H. Stiehl, and H. T. Witt (1967) in whole cells of *Chlorella,* and by G. Döring, J. Bailey, W. Kreutz, and H. T. Witt (1968) in "system II particles" (see Chapter 16) obtained from spinach. Using repetitive fast flashes (of $\sim 10^{-6}$ sec duration), they were able to observe absorbancy changes decaying with a half-time of about 2×10^{-4} sec (in contrast to changes in P700 that decay with a half-time of about 2×10^{-2} sec). The peak of this change in whole cells is at 690 nm; but in separated system II particles, it appears to be shifted to about 682 nm; it is absent in

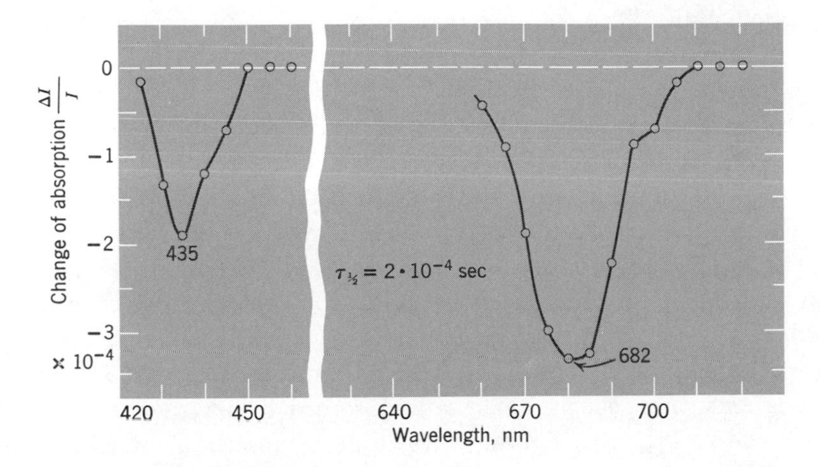

FIG. 14.5 Absorption changes with a lifetime of ~2.10^{-4} sec in system II particles from spinach. Electron acceptor: K_3 [$Fe(CN)_6$]5.10^{-5} M/1. Excitation: 620–710 nm (left curve) or 380–550 nm (right curve). Repetitive pulse technique has been used. For each measuring point, 4096 flashes were fired. (G. Döring, J. L. Bailey, W. Kreutz, and H. T. Witt, 1968.)

system I particles. The corresponding Soret band is observed at 435 nm (Fig. 14.5).

PLASTOQUINONE AND PLASTOCYANIN

In addition to cytochromes, P700, and P690, two other catalytic compounds were found to contribute to difference spectra of chloroplasts. One is called *plastoquinone.* It has an E_0' of about 0 volt. The other is a protein called *plastocyanin;* it is blue because it carries a copper atom. Its potential (E_0') is +0.36 volt. It seems plausible that both plastocyanin and plastoquinone act as intermediate redox catalysts in the electron path from system II to system I, plastoquinone close to cytochrome b_3, and plastocyanin close to cytochrome f (Fig. 14.4).

Plastoquinone (*PQ*), is one of several quinones found by fractionation of the photosynthetic system. The first suggestion that one of them is involved as intermediate in photosynthesis came from the antagonistic effect of light absorbed in pigments systems I and II on its oxidation-reduction state. When the spectrum of oxidized PQ is measured against that of the corresponding hydroquinone, a difference band is found in the ultraviolet at 254 nm. In H. T. Witt's laboratory in Germany (1962), absorption changes in illuminated algae were noted in this region (see Fig. 19.2*d*). In the green alga *Chlorella,* light in the region 650–660 nm (absorbed predominantly in PSII) caused a decrease in absorption at 254 nm (that is, reduction of PQ); while far-red light (absorbed in PSI), caused an increase in absorption (that is, oxidation of PQH_2). In Duysens' laboratory in Leyden (Holland), Amesz (1964) provided similar evidence for the role of plastoquinone in the blue-green alga *Anacystis* (maximum absorbance change is, in this case, at 255–260 nm). In this alga, orange light (620 nm), absorbed by phycocyanin, caused reduction of PQ, and red light (680 nm), absorbed by Chl *a,* its reoxidation.

Compared to the cytochromes, plastoquinones are present in the algae in about ten times larger quantities.

Plastocyanin (*PC*) is a copper protein discovered in the green alga *Chlorella* by S. Katoh (1960) in Japan. D. C. Fork and co-workers at the Carnegie Institution at Stanford observed that light absorbed in pigment system I oxidizes plastocyanin (as shown in Fig. 14.9*c* by changes in optical density at 590 nm, the peak of the difference band between the spectra of oxidized and reduced plastocyanin). Light absorbed in pigment system II accelerated the reverse reaction—the reduction of oxidized plastocyanin. From some poisoning experiments (a method much used in enzymology), Fork concluded that plastocyanin precedes cytochrome *f* in the electron transport chain (in the green alga *Ulva,* and several other organisms). The oxidation-reduction potential of plastocyanin ($E_0' = +0.38$) is, in fact, slightly above that of cytochrome *f*. However, the exact sequence is not yet certain.

BACTERIA

Difference spectra have revealed that photosynthetic units in photosynthetic bacteria also contain "energy traps" in the form of specialized bacteriochlorophyll molecules. The maxima of the negative difference bands are at 840, 870 or 890 nm, depending on the organism. This can be attributed to photooxidation of a "trap" (similar to P700 in green algae). Absorbancy changes attributable to bacterial cytochromes, and quinones of a certain type (ubiquinones, related to the plastoquinone) also have been noted in bacterial systems. Action spectra for the oxidation of various cytochromes in bacteria suggest the presence of two pigment systems and two light reactions.

Chapter 15

Fluorescence and the Two Pigment Systems

Preceding chapters dealt with experimental evidence underlying the concept of two primary photochemical reactions sensitized by two separate pigment systems, being involved in photosynthesis. Further arguments supporting this concept, and supplying additional details, can be derived from the study of chlorophyll fluorescence in vivo.

Fluorescence measurements permit practically instantaneous nondestructive monitoring of rapidly changing chemical systems that involve fluorescent reactants, intermediates, or products. Photosynthesis is such a process. Chlorophyll is fluorescent in vivo as well as in vitro, and so are the phycobilins. For reasons to be explained later, we are mainly interested in the fluorescence of chlorophyll.

The fluorescence yield of chlorophyll *a* (ratio of the number of quanta emitted and the number of quanta absorbed) is much lower in vivo than in vitro (3–6% in vivo, against 30% in vitro). For a time, the very existence of chlorophyll fluorescence in plants was disputed. However, modern experimental techniques make it possible not only to demonstrate the existance, and measure precisely the intensity of plant fluorescence, but also to determine its spectral composition and the action spectrum for its excitation. A scheme of such an instrument is shown in Fig. 15.1.

It appears that several forms of chlorophyll *a* contribute to fluores-

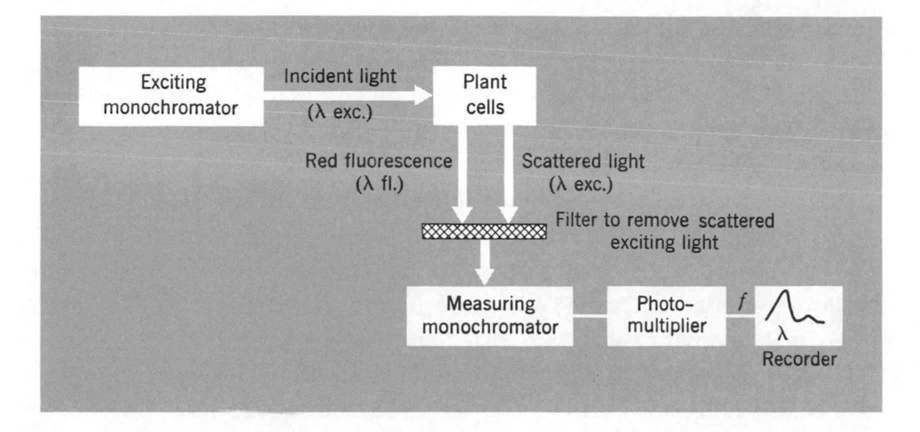

FIG. 15.1 Schematic diagram of a spectrofluorometer. A common modification of this setup is to collect fluorescence from the same side where the incident light hits the cuvette containing plant cells; this is done to reduce the reabsorption of fluorescence.

cence in vivo (as they were shown in Chapters 9 and 13 to contribute to light absorption). Each has its own characteristic emission spectrum. The yield of fluorescence of these components is quite different; and their excitation spectra differ, too. They have a different dependence on intensity of illumination, and change differently during the induction period.

We have already discussed some applications of fluorescence studies in Chapter 12, when talking of excitation energy transfer between pigments in vivo. These studies have revealed that excitation energy is transferred from all accessory pigments to chlorophyll *a*. The transfer from the phycobilins to chlorophyll requires a few nanoseconds; this gives enough time for phycoerythrin and phycocyanin to emit some fluorescence. Energy transfer from chlorophyll *b* to cholorophyll *a* is so fast that no Chl *b* fluorescence has been observed in plants.

The yield of fluorescence of the phycobilins proves to be independent of the intensity of illumination, presence of poisons, and other factors affecting the rate of photosynthesis. The yield of fluorescence of chlorophyll *a*, on the other hand, in addition to being much smaller than in solution, is significantly affected by these factors. Clearly, chlorophyll *a* fluorescence competes with the use of excitation energy for photosynthesis; while the fluorescence yield of the phycobilins is limited by processes—such as energy transfer and internal energy conversion—whose efficiency does not depend on the occurrence and rate of the photochemical process. This is why we are interested in Chl *a* fluorescence much more than in that of the phycobilins.

The *emission spectrum* of chlorophyll *a* in vivo has a main band, with a maximum at 685($\pm$2) nm, and a vibrational "satellite" band at about 740 nm (Fig. 15.2). These bands originate in transitions from the lowest excited electronic state of chlorophyll *a* to its ground level, either in the nonvibrating state (0,0 band) or in its first vibrating state (0,1 band). The size of the vibrational quantum corresponding to the wavelength difference between 685 and 740 nm (about 1460 cm^{-1}) sug-

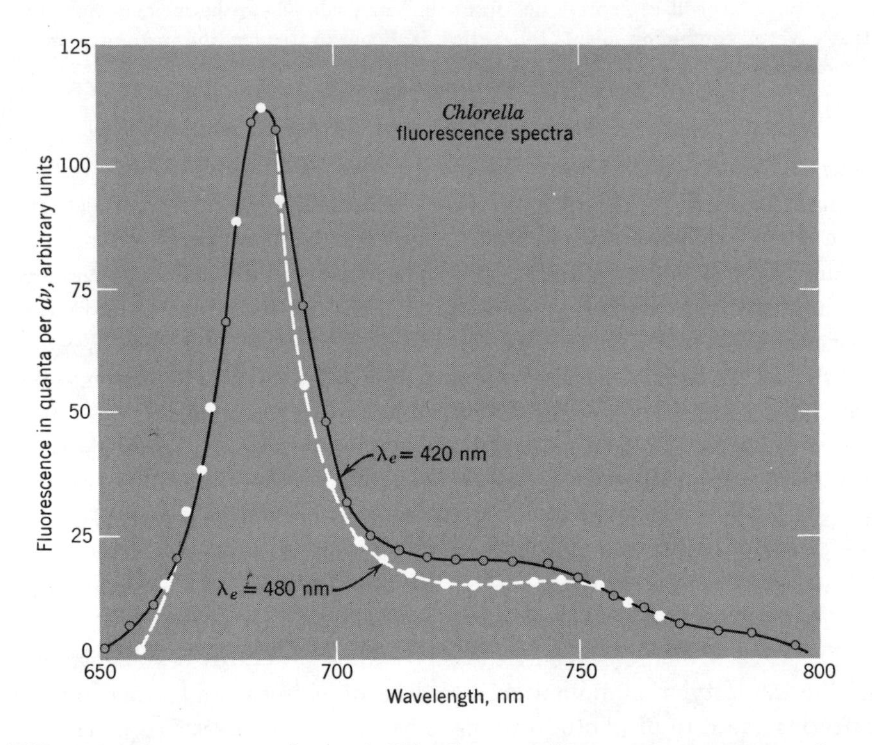

FIG. 15.2 Fluorescence spectra of *Chlorella pyrenoidosa* excited at 420 nm (λ_e = 420) and 480 nm (λ_e = 480), showing increased excitation of F720 by light preferentially absorbed in Chl *a* (420 nm) compared to light preferentially absorbed in Chl *b* (480 nm). (Duysens, 1952; confirmed in spinach chloroplasts by Govindjee and L. Yang, 1966.)

gests that the vibration coupled with the electronic transition may be a C—H "bond deformation" vibration—a fact of some interest in relation to the mechanism by which excitation energy is used for the primary photochemical process.

Excitation in the blue-violet (Soret) band leads to a higher, "blue," excited electronic state; but it is quickly followed (in vivo as well as in vitro), by radiationless internal conversion into the "red" excited state, with the loss of about one-third of excitation energy. This transition is so fast ($<10^{-10}$ sec), that no emission originating in the "blue" state has been observed.

E. C. Wassink (in Holland), and James Franck (in Chicago), had observed a striking increase (approximate doubling) of the fluorescence yield of chlorophyll a in vivo with increasing intensity of illumination, apparently associated with the light saturation of photosynthesis. Hans Kautsky and co-workers (in Germany) first described another striking phenomenon—changes in fluorescence intensity that occur when cells are suddenly brought from darkness into light, during the so-called *induction* period of photosynthesis. Application of more sensitive instruments has revealed that both these changes affect differently the several spectroscopic components of fluorescence.

Studies in liquid nitrogen and liquid helium showed considerable increase in the yield of fluorescence at low temperatures, and appearance of new emission bands probably due to chlorophyll *a* forms not present (or nonfluorescent) at room temperature.

Studies with plane-polarized light provided some information about the spatial arrangement of chlorophyll molecules in the photosynthetic units.

We will discuss below only a few results of all these studies, most clearly related to tht two-pigment-systems hypothesis.

SPECTRUM OF CHLOROPHYLL *a* FLUORESCENCE IN VIVO

Fluorescence at Room Temperature

"STRONGLY" AND "WEAKLY" FLUORESCENT CHL *a* FORMS IN VIVO. The main emission band of Chl *a* in vivo (F685)[1] appears to contain, at

[1] We designate fluorescence bands by the letter F; absorption bands can be similarly designated by the letter A.

room temperature, a weak component, at (or beyond) 700 nm. (Its precise position is obscured by the above-mentioned vibrational subband at 740 nm.) The doublet nature of F685 is revealed, for example, by matrix analysis. Exciting fluorescence at one set of wavelengths, λ_{ex}, and measuring its intensity at another set of wavelengths, λ_{em}, a series of matrices can be constructed. If all 2×2 matrices[2] are not equal to zero, there must be, according to Gregorio Weber, at least two fluorescence emitters; if some 3×3 matrices do not disappear, there must be at least three fluorescence emitters, etc. Applying Weber's analysis to the fluorescence of spinach chloroplasts, Louisa Yang and Govindjee found that at room temperature (at least) two Chl *a* forms contribute to fluorescence. S. S. Brody obtained similar results on *Euglena*.

When *Chlorella* cells (or spinach chloroplasts) are excited at 440 nm (that is, with light absorbed preferentially by chlorophyll *a*), the long-wave component (F720) is (relatively) stronger than when 480 nm light (absorbed preferentially by chlorophyll *b*) is used for excitation (Fig. 15.2). This suggests that a larger part of F720 originates in PSI. (We have concluded before that this system contains less chlorophyll *b* than PSII.) The main band, F685, probably originates in PSII, which contains a larger part of Chl *b*.

The conclusion that Chl a_{II} (Chl *a* in system PSII) is the main source of chlorophyll *a* fluorescence in vivo is confirmed by a variety of observations. L. N. M. Duysens (in Holland) and C. Stacy French (in Stanford, California) have observed that in red algae, green light, absorbed mainly by the phycobilins, is—paradoxically!—more efficient in exciting chlorophyll *a* fluorescence than light absorbed in chlorophyll *a* itself. (These were the observations that first led Duysens to postulate the existence, in red algae, of two forms of chlorophyll *a*, the more strongly fluorescent one being preferentially associated with the phycobilins.) The same is true of blue-green algae (Fig. 15.3). (It was suggested in Chapter 13 that phycobilins are more abundant in PSII than in PSI!)

We can thus assume that the "fluorescent" form of Chl *a* belongs

[2] A 2×2 matrix consists of four numbers—in our case, of four intensity values, $I\lambda_{ex}\lambda_{em}$, corresponding to excitation wavelengths λ_{ex_1} and λ_{ex_2}, and emission wavelengths λ_{em_1} and λ_{em_2}. The value of the matrix is the difference $I(\lambda_{em_1}\lambda_{em_1})I(\lambda_{ex_2}\lambda_{em_2}) - I(\lambda_{ex_2}\lambda_{em_1})I(\lambda_{ex_1}\lambda_{em_2})$.

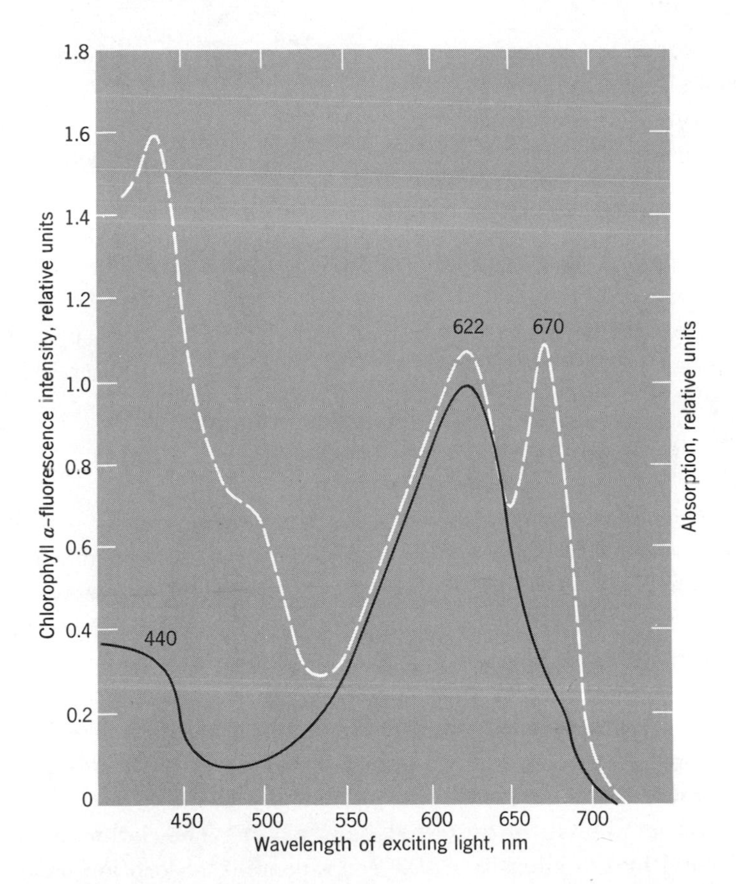

FIG. 15.3 Fluorescence excitation spectrum of blue-green algae (*Anacystis nidulans*) measured at 745 nm, compared to absorption spectrum (dashed curve), indicating more efficient excitation of Chl *a* fluorescence by light absorbed in phycocyanin (at 620 nm) than by light absorbed by Chl *a* itself (at 420 and 670 nm). (G. Papageorgiou and Govindjee, 1967.)

to PSII, and the "nonfluorescent" (or, more exactly, less fluorescent form), to PSI.

Another indication of the existence of two forms of Chl *a*, with different fluorescence yields, was obtained from measurements of the average lifetime, τ of chlorophyll *a* fluorescence. It was found to be about 0.7 nsec in weak light (W. J. Nicholson and J. Fortoul, I.B.M., New York; G. Singhal, University of Illinois)*, and 2 nsec in strong light (R. Lumry

* About 0.6 nsec according to W. J. Nicholson and J. Fortoul (1967) and down to 0.4 nsec, according to A. Muller, R. Lumry and M. S. Walker (1969).

and co-workers, University of Minnesota), as compared to 4.5 nsec. in chlorophyll solution in ether (in weak or strong light). If one compares the lifetime ratios between Chlorella cell and ether solution (6.5 in weak light, and 2.5—based on 1.3 nsec value— in strong light) with the ratios of the fluorescence *yields* in vitro and in vivo (about 10:1 in weak light and 5:1 in strong light), one finds a discrepancy, which suggests that only part of Chl *a* in vivo is fluorescent. The argument is as follows:

If the excitation of a molecule can end either by emission of fluorescence (yield, Φ) or by an internal quenching process ("internal conversion," yield, $1 - \Phi$), the actual, directly observable lifetime of excitation will be

$$\tau = \Phi\tau_0 \tag{15.1}$$

—a relation that can be easily derived from consideration of two competing processes, fluorescence emission and internal conversion, treated as two independent "first order" processes (that is, processes the rate of which is proportional, at any given time, to the number of excited molecules present). The constant τ_0 is called "natural" lifetime; it is the lifetime corresponding to $\Phi = 1$, that is, to the case when excitation is limited by fluorescence emission only. Natural lifetime must be inversely proportional to the probability of transition from the excited to the ground state, and thus also to that of transition from the ground state to the excited state (see Chapter 10). The latter probability is measured by the intensity of the corresponding absorption band (more precisely, by the integral of the plot $\alpha = f(\lambda)$, where α is the absorption coefficient at wavelength λ, integration being extended over the whole absorption band). This integration gives, in the case of Chl *a* in vitro, $\tau_0 = 15$ nsec. Inserting this value, and the above-mentioned experimental value of τ (4.5 nsec) into Eq. 15.1, we obtain

$$\Phi = \frac{\tau}{\tau_0} = \frac{4.5}{15} = 0.3, \text{ or } 30\% \tag{15.2}$$

—in good agreement with direct determination of the fluorescence yield of Chl *a* in solution. However, no such agreement is found in the case of fluorescence in vivo. Since it is difficult to determine the integral of the absorption band in vivo, one makes, in this calculation, the plausible assumption that the natural lifetime of excited chlorophyll, τ_0, is the same in vivo as in vitro. With $\tau = 0.7$ nsec (weak light) or 1.3 nsec (strong light) one calculates from Eq. 15.2, $\Phi = \tau/\tau_0 = 5\%$ (weak light) or 9% (strong light), while direct measurements of the fluorescence yield in vivo give 3% in weak light and 6% in strong light. This discrepancy can be explained by assuming that a significant fraction of Chl *a* in vivo is nonfluorescent (or only weakly fluorescent). In this case, Φ should be determined by referring the emission not to *total* chlorophyll absorption, but only to absorption by the fluorescent component. The results suggest that about one half of chlorophyll *a* in vivo is nonfluorescent, in weak as well as in strong light.

ACTION SPECTRUM OF CHL *a* FLUORESCENCE IN VIVO: THE "RED DROP." If the "weakly fluorescent" form of chlorophyll *a* in vivo absorbs relatively more strongly on the long-wave side of the red absorption band, a "red drop" can be expected in the action spectrum (the plot of the quantum yield of fluorescence, Φ, as function of wavelength of exciting light, λ_{ex}). Duysens (1952) did in fact note such a decline in green algae beyond 680 nm. Direct measurements of Φ (as well as calculations based on relationship between the absorption and the emission spectrum) made in Urbana, also indicated a drop of Φ, beginning (in *Chlorella*) at about 675–680 nm (see Fig. 15.4).

The action spectrum of chlorophyll fluorescence in red algae is

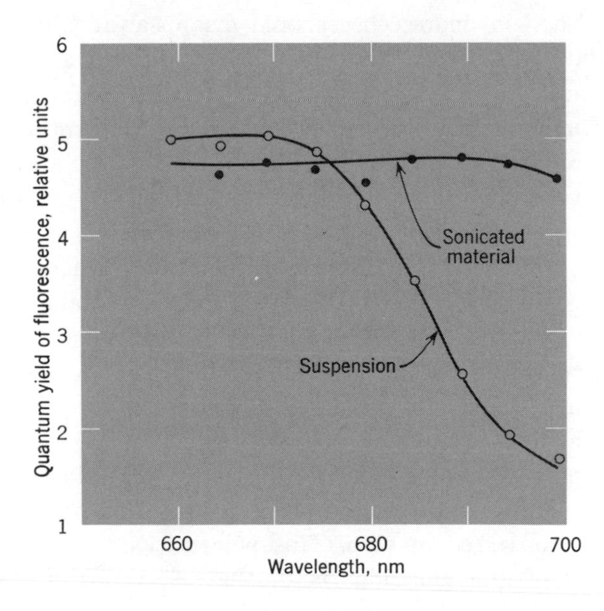

FIG. 15.4 Red drop in action spectrum of fluorescence in *Chlorella* and its disappearance in aerobic acid sonicates of *Chlorella*. (M. Das and Govindjee, 1967.)

quite different from that in green algae, because almost all Chl *a* seems to be present there in PSI. Consequently, the spectral region of predominant absorption in PSI begins, in red algae, at much shorter wavelengths than in the green ones, and the "red drop" starts correspondingly earlier, when absorption in phycocyanin ceases to be strong compared to that in Chl *a*. (This consideration applies to the action spectrum of photosynthesis as well as to that of fluorescence.)

In the red alga *Porphyridium cruentum,* and the blue-green alga *Anacystis nidulans,* the fluorescence yield is found, in fact, to decline already at 640–660 nm.

One is tempted at first to associate the "weakly fluorescent" component of Chl *a* with Chl *a* 680, as identified from absorption data (see Fig. 9.7). However, quantitative analysis argues against this identification. Instead, the "red drop" seems to be associated with a third, minor Chl *a* component present only in PSI (French's "Chl *a* 695," or Butler's "C700"), while both main forms, Chl *a* 670 *and* Chl *a* 680, present in both PSI and PSII, contribute about equally to fluorescence. The presence of Chl *a* 695 in PSI makes *all* Chl *a* in this system non- (or weakly) fluorescent, by draining and dissipating the excitation energy from the two main components.

The "red drop" in fluorescence should disappear if Chl *a* 695 could be preferentially destroyed. Under aerobic conditions, and at low pH, prolonged sonication[3] of *Chlorella* cells lead to such preferential destruction. This is indicated by difference between the absorption spectra of "sonicates" prepared in alkaline medium (pH 7.8) in absence of air, and of sonicates prepared in acid medium (pH 4.5) in presence of air. The latter appear deficient in Chl *a* 695 and show, as expected, no "red drop" in the action spectrum of their fluorescence (Fig 15.4).

POLARIZATION OF CHL *a* FLUORESCENCE. We suggested above that Chl *a* 695 is a component of PSI. Polarization experiments suggested that it may be a more orderly form of chlorophyll *a* than the main components, Chl *a* 670 and Chl *a* 680. In Chapter 12, we said that Chl *a* fluorescence in vivo, excited by polarized light, is largely depolarized; this observation was used as evidence of energy migration in the photosynthetic unit. However, a certain weak polarization of Chl *a* fluorescence has been observed in vivo. This polarization is strongest in the long-wave end of the emission band, that is, probably to emission from PSI.

[3] "Sonication" means treatment by ultrasonic waves.

INDUCTION OF CHL *a* FLUORESCENCE IN VIVO. The complexity of Chl *a* fluorescence in vivo is confirmed by its variation with time during the induction period of photosynthesis. When *Chlorella* cells are exposed to bright light after a long dark period, the fluorescence yield Φ practically instantaneously reaches a certain level [*O* in Fig. 15.5*a*]. Within a

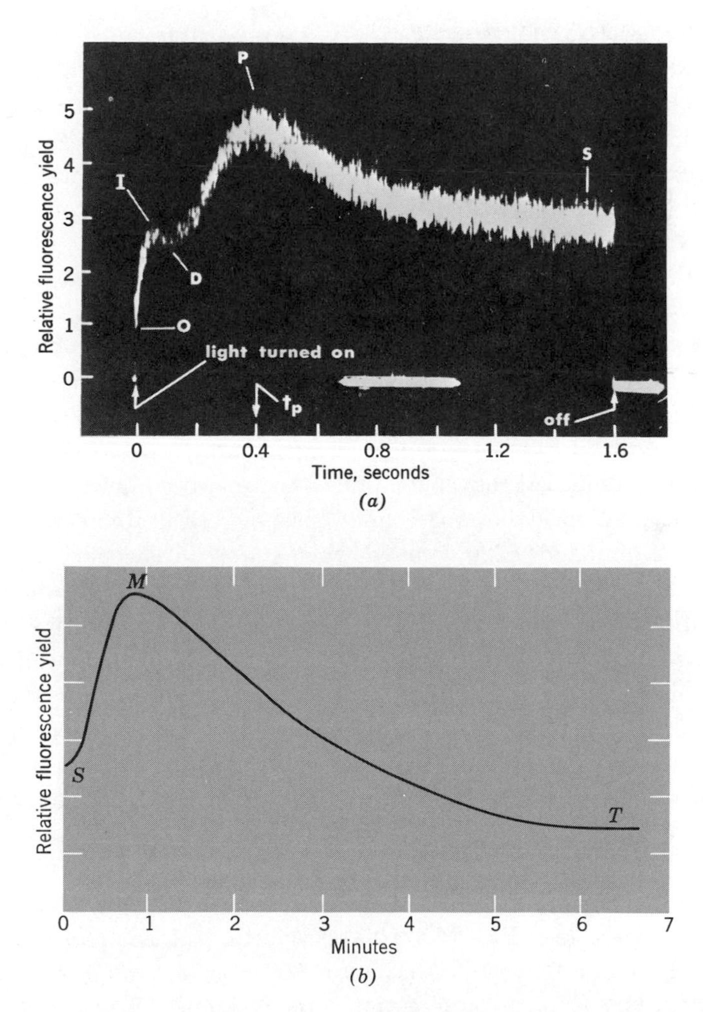

FIG. 15.5 Fluorescence induction in aerobic *Chlorella pyrenoidosa*. Wavelength of measurement: 685 nm; excitation: blue light. (*a*) Picture taken on the oscilloscope screen—up to 2 seconds. (J. C. Munday and Govindjee, 1969.) (*b*) Measurements with a Brown recorder—from 3 seconds to 6 minutes. (G. Papageorgiou and Govindjee, 1968.) For meaning of letters, see text.

few milliseconds, Φ rises to a higher level, *I*; it then decreases slightly to *D*, increases again to a peak *P*, reached after about 0.5 seconds, and declines to a more or less steady state *S*. A slow rise to a second maximum (*M*), follows, which takes about 40 seconds, and is followed by a decline to a terminal stationary state (T), reached after several minutes (Fig. 15.5*b*).

Jean Lavorel (at Gif-sur-Yvette, in France) first noted that at time *O* fluorescence contains relatively more of the weak, long-wave fluorescence band (F720) than in the peak *P*. These experiments, and many others, including those in our laboratory, suggest that it is the main fluorescence band at 685 nm (which we presume to originate in system II) that accounts for the induction phenomena, while the weak fluorescence from system I remains practically constant.

Fluorescence at Low Temperatures

Emission spectra. Sharpening of emission bands at low temperature (caused by cessation of intramolecular vibrations) is a useful means to reduce their overlapping. The quantum yield of total fluorescence is considerably increased at low temperatures, since both external quenching and internal conversion processes are slowed down.

S. Brody, in our laboratory, first observed (Fig. 15.6) that a new strong emission band, F720, appears when *Chlorella* is cooled to —190°C (77°K). Several investigators in our laboratory and elsewhere have observed another long-wave band, F695. An extension of this work to liquid helium temperature (4°K) by Frederick Cho in our laboratory confirmed the existence, at that temperature, of three emission bands, F689, F698, and F725. The overall shape of the emmission spectrum exhibits a strong temperature dependence, caused by changes in relative intensities of these three components. The F698 band appears only between 4°K and 140°K (Fig. 15.7).

Excitation of chlorophyll *b* (in green plants) or of the phycobilins (in red or blue-green algae) leads to a weaker F720 and stronger F685 and F695 bands (see Fig. 15.8) than excitation of chlorophyll *a*. This suggests that not only F685 but also F695 belongs mainly to system II, while F720 belongs preferentially to system I. This is confirmed by the finding that particles (prepared by the N. K. Boardman and J. Anderson's digitonin fractionation method, see Chapter 16) which perform preferentially light reaction I, show, at 77°K, a preponderance of

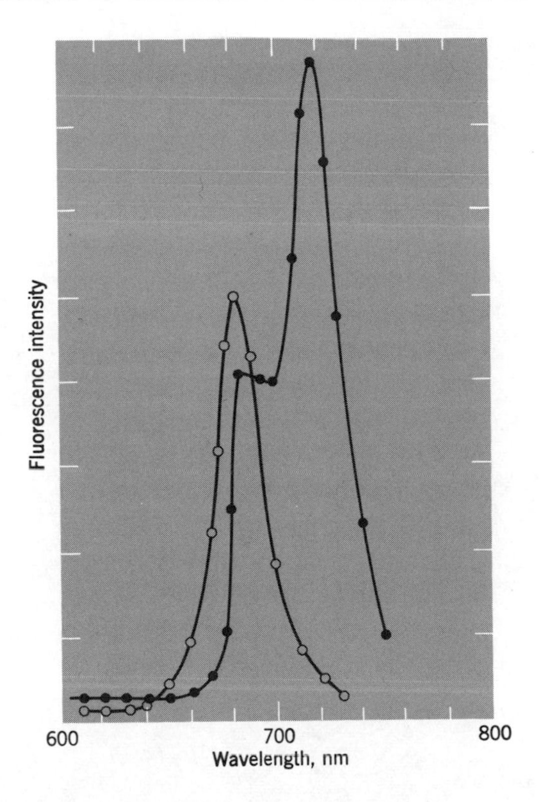

FIG. 15.6 Fluorescence spectrum of *Chlorella* at −177°C, showing a strong band at 720–725 nm (black dots); spectrum at room temperature is shown for comparison (circles). (S. Brody, 1958.)

F725, whereas particles that perform preferentially light reaction II, appear enriched in F685 and F695-emitting material.

FLUORESCENCE FROM "TRAPS"? The energy trap in pigment system II has been only recently identified by difference spectroscopy; the rate of regeneration of this trap is very fast ($\sim 10^{-4}$ sec) as opposed to that of P700 ($\sim 10^{-2}$ sec) (see Chapter 14). Cooling should favor fluorescence emission from traps, because at low temperature, energy quanta caught in a trap will find it difficult to get back into the "bulk" of the pigment before emitting fluorescence. The trap depth, estimated from the displacement of the absorption bands (for example, the shift from 680 nm, where the absorption band of the "bulk" is located, to 690 nm, where we sup-

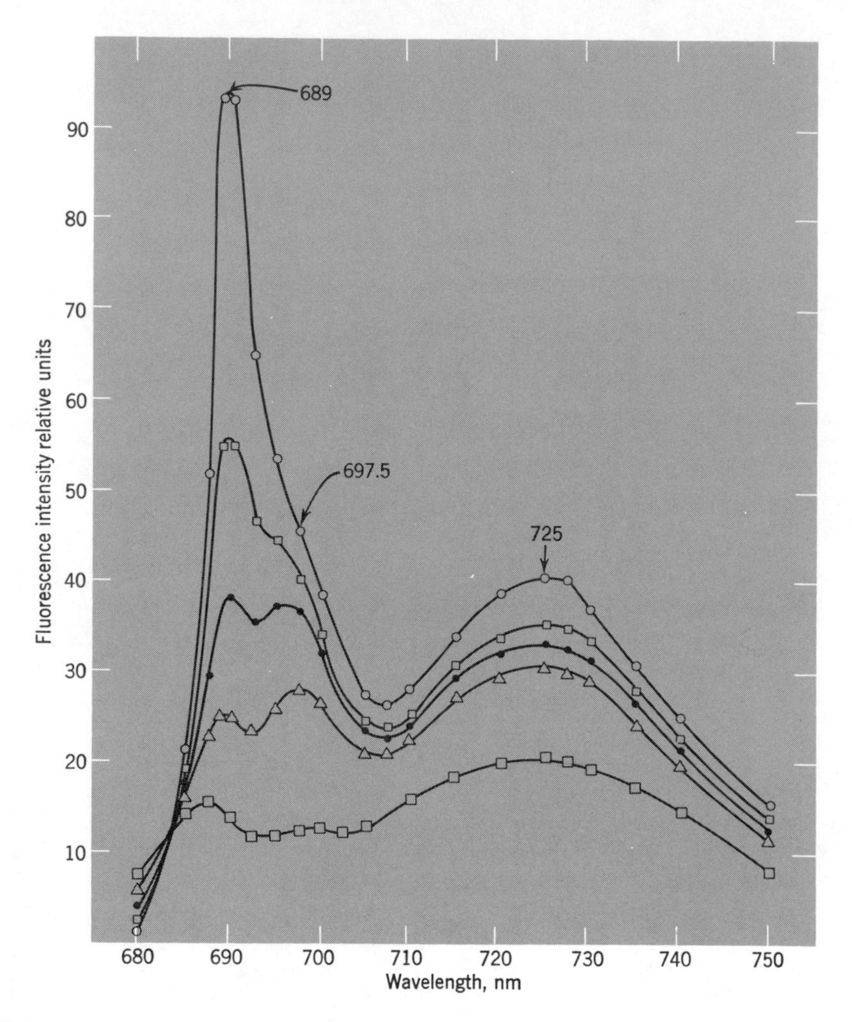

FIG. 15.7 Fluorescence spectra of *Chlorella* at several temperatures down to —269°C. (□, —196°C, lowest curve; Δ, —218°C; ●, —233°C; □, —247°C; ○,—269°C.) (After F. Cho, J. Spencer, and Govindjee, 1966.)

pose the band of the "trap" in PSII is found), is 1–2 kT at room temperature, 5–8 kT at 77°K (in liquid nitrogen) and 100–150 kT at 4°K (in liquid helium); the probability of escaping from this trap in $< 10^{-8}$ sec. (the lifetime of excitation), is very high in the first case and practically nil in the last one. The F695 band, which appears only

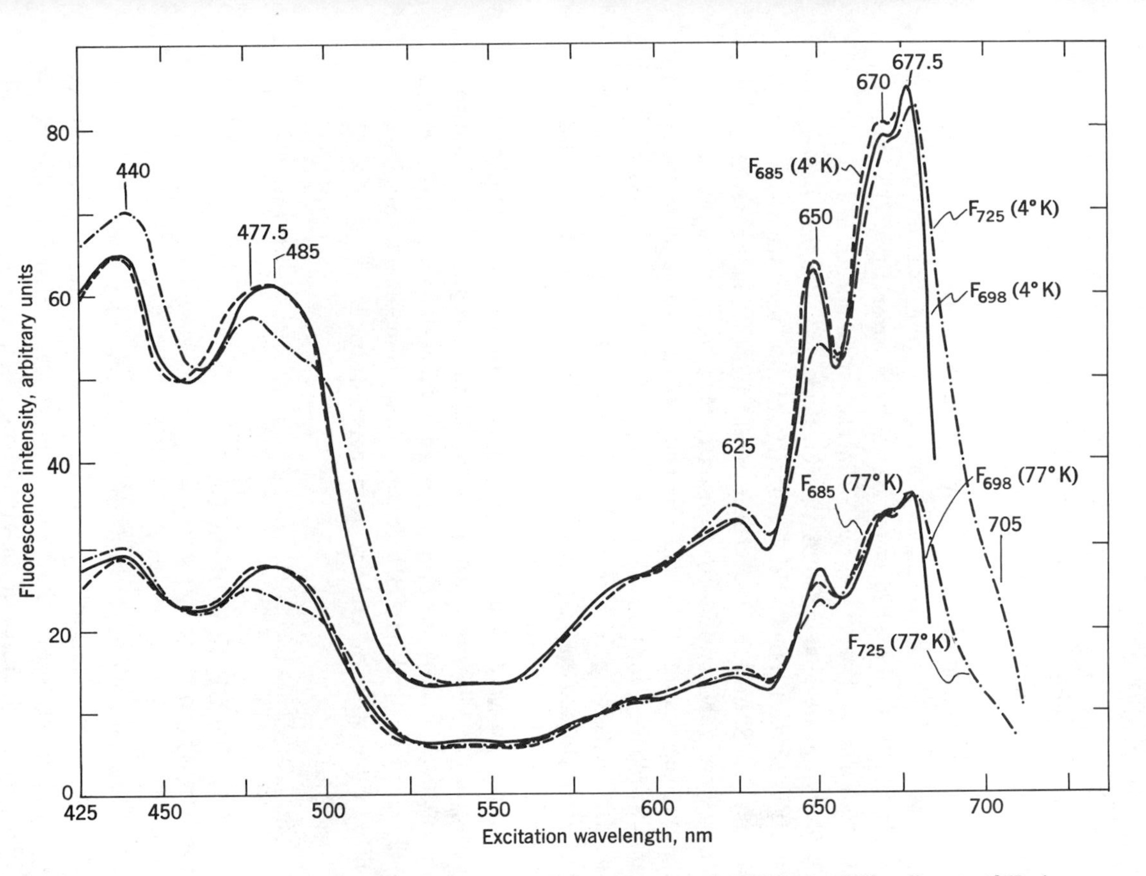

FIG. 15.8 Excitation spectra of fluorescence (for F685, F698, and F725) in *Chlorella* at 4°K (upper curves) and 77°K (lower curves). (F. Cho and Govindjee, 1969.)

below 140°K, is thus a likely candidate for fluorescence emission from the trap in system II.

ACTION SPECTRA. That the bands F685 and F698 are excited mainly by absorption in system II, and the band F725 mainly by absorption in system I, is confirmed by measurements, by F. Cho, of the excitation spectra of these bands (in *Chlorella*) at 4°K (Fig. 15.8). For example, we note that the ratio of the intensity of fluorescence excited at 440 nm (in the peak of Chl *a* absorption band) to that excited at 485 nm (in the peak of Chl *b* absorption band) is greater for F725 than for F685 and F698. Furthermore, the F725 excitation spectrum has a much less pronounced peak at 650 nm (in the peak of Chl *b* red band) than those of F685 and F698. Near identity of the excitation spectra for F685 and F698 in Fig. 15.8 confirm that both belong to the same system (system II) while the presence of a band >700 nm in the excitation spectrum of F725 confirms that the latter originates mainly in system I (probably, in Chl *a* 695).

Another interesting feature of Fig. 15.8 is the appearance, at 4°K, of two separate excitation peaks in the red, at about 670 and 680 nm. These wavelengths clearly correspond to the two absorption peaks of chlorophyll *a* in vivo identified by analysis of the red band at room temperature (see Chapter 9). Appearance of these peaks in the excitation spectra of both F725 (Chl a_{I}) and F698 (Chl a_{II}) confirms the suggestion that PSI as well as PSII contain both components, although apparently in somewhat different relative concentration.

DEPENDENCE OF FLUORESCENCE YIELD ON LIGHT INTENSITY

In dye *solutions* (including those of chlorophyll) the fluorescence intensity (F) is proportional, within wide limits, to the intensity of illumination (I); in other words, the *yield* of fluorescence does not depend on it. Chlorophyll *a* fluorescence in vivo shows, however, a very significant dependence of the yield on the intensity of illumination (see Fig. 15.9, where the upper curve shows the fluorescence *intensity, F*, and the lower curve, the fluorescence *yield, F/I* as function of the intensity of illumination, I). The second curve illustrates the fact, mentioned earlier in this

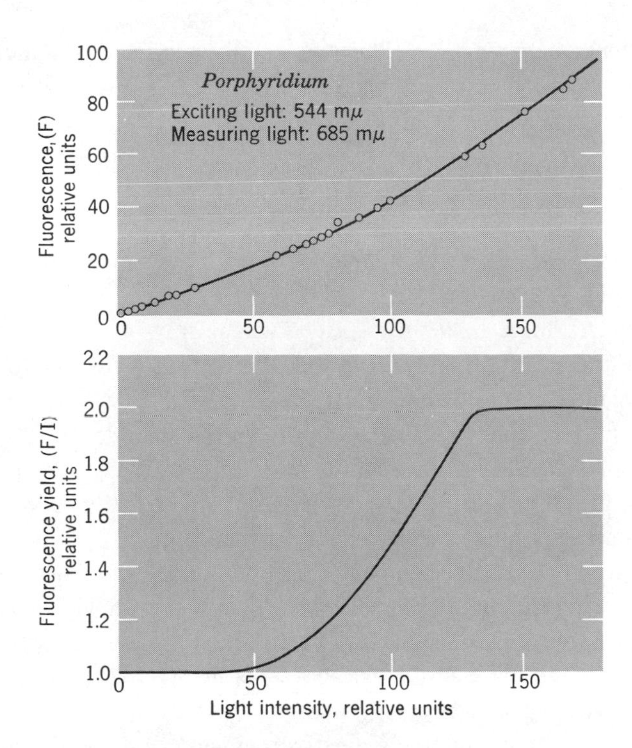

FIG. 15.9 Fluorescence intensity (F) and fluorescence yield (F/I) of *Porphyridium* as function of intensity of illumination. (A. Krey and Govindjee, 1964.)

chapter, that in light strong enough to saturate photosynthesis, the yield of fluorescence approximately doubles (rising from about 0.03 to about 0.06).

James Franck offered an interesting explanation for the doubling of the fluorescence yield at high light intensities. He suggested that, in photosynthesis, two primary photoreactions are brought about by chlorophyll molecules (or reaction centers in a photosynthetic unit), acting in the short-lived, fluorescent *singlet* state, and in the long-lived *metastable,* nonfluorescent *triplet* state, respectively. In the first case, photochemical action competes with fluorescence; in the second case, there is no such competition, because fluorescence is emitted before the molecules are transferred into the triplet state. The reason for this difference can be that the first reaction occurs practically instantaneously inside the "pigment complex," while the second requires a kinetic encounter with an independent reactant. The yield of fluorescence does not depend, in the second case, on whether the triplet molecules are using their energy for a photochemical reaction, or losing it in some other way. During efficient photosynthesis, an equal number of reaction centers must operate in the singlet and in the triplet state. In low light, all fluorescence originates in the Chl *a* molecules (or photosynthetic units) that operate in the triplet state; while the fluorescence of molecules (or units) that operate in the singlet state, is completely quenched by photochemical competition. At high light intensities, when only a small fraction of the total number of quanta absorbed are utilized for photosynthesis, this utilization ceases to compete significantly with fluorescence emission. Consequently, in strong light *all* excited photosynthetic units contribute to fluorescence—while in weak light, only one half of them participate in it. This explains, according to Franck, why the quantum yield of fluorescence in strong light is twice that in weak light.

Franck proposed this ingenious explanation in the framework of a theory according to which the "triplet" and the "singlet" sensitization occurs alternately in one and the same Chl *a* molecule (or one and the same photosynthetic unit), while we have arrived, in the preceding chapters, at a concept of two separate photosynthetic systems sensitizing two successive primary photochemical processes. One could suggest that Franck's concept should be transferred to this picture, by assuming that one of the two systems operates in the singlet and the other in the triplet state; but closer analysis of the spectral composition of fluorescence in weak and in strong light do not support this hypothesis. Spectral analysis suggests that PSI contributes a weak but constant (intensity-independent) amount of fluorescence (F720) while PSII, which contributes most of the fluorescence in weak light (mainly F685) doubles its yield in strong light. In the red alga *Porphyridium cruentum*, a doubling of the quantum yield of fluorescence in strong light is observed when light is absorbed in the red pigment phycoerythrin (that belongs mainly to PSII); while no such effect is observed when excitation takes place in chlorophyll *a*, that is, predominantly in PSI (Fig. 15.10). The same effect as increased light intensity is caused by poisoning PSII selectively with poisons such as DCMU, [3- (3′, 4′ dichlorophenyl) 1,1-dimethyl urea]. Here, too, the weak PSI fluorescence remains unaffected, while the strong PSII fluorescence doubles or even trebles in intensity.

If Franck's explanation of the rise of Φ in strong light as resulting

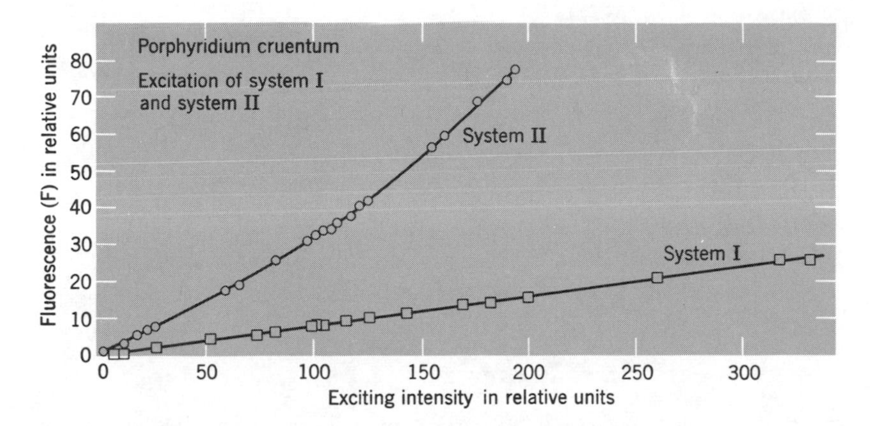

FIG. 15.10 Fluorescence intensity of *Porphyridium* as function of intensity of illumination for absorption in PSII (system II) and PSI (system I). (A. Krey and Govindjee, 1966.)

from transition from "one system fluorescence" to "two systems fluorescence" is abandoned, the doubling of the yield must be considered as a rather accidental result of a change, in strong light, of the relative rates of fluorescence emission and fluorescence quenching processes within system II.

THE TWO-LIGHT EFFECT

In 1960, we observed that chlorophyll *a* fluorescence in *Chlorella,* excited with red or blue light (absorbed in both PSI and PSII) became weaker when far-red light (absorbed mainly in PSI) was added. At the Agricultural Experiment Station at Beltsville, Md., Warren Butler found, in 1962, that preillumination with a strong orange-red light ("system II light"), causes, in *Chlorella,* enhanced fluorescence, while preillumination with strong far-red light ("system I light") weakened it. In Leiden (Holland), Duysens and Sweers (1963) superimposed a weak modulated light beam on a strong constant background light, and measured the fluorescence caused by the modulated beam. The "two-light effect" could be easily confirmed in this way (Fig. 15.11).

H. Kautsky and co-workers (1960) suggested that their fluorescence

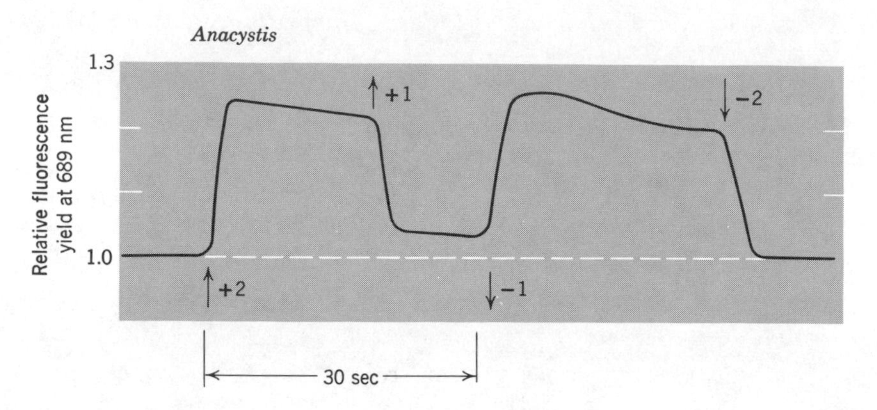

FIG. 15.11 Two-light effect on fluorescence of *Anacystis;* +2, addition of system II light; and +1, addition of system I light. (L. N. M. Duysens and H. E. Sweers, 1963.)

induction data could be best explained by postulating two light reactions in photosynthesis. Duysens proposed that, in PSII, electrons taken from water reduce an intermediate catalyst which he called Q. (The symbol Q stands for "quencher" of fluorescence.) The increase in fluorescence yield, caused by strong light absorbed in pigment system II, is due, he suggested, to accumulation of reduced (Q^-), which cannot accept electrons and is therefore a "nonquencher." When light absorbed in PSI is added, Q^- is reoxidized to Q, and fluorescence decreases.

Detailed studies of changes in PSII have been carried out in several other laboratories (including those at the University of Illinois, Kautsky's in Germany, Bessel Kok's at Baltimore, Jean Lavorel's and Pierre Joliot's in France), by observing changes in fluorescence during the induction period. These experiments provide the closest approach, so far, to the inner sanctum of photosynthesis—that is, to events that occur upon illumination in the photosynthetic pigment complex itself; but their complexity makes it impossible to describe them in detail in this introductory text.

Interpretations, suggested by several groups of observers, were all based on Fig. 14.4 according to which light absorption in PSII causes the reduction of a "primary oxidant" (labeled Y in the figure; Duysens' Q), which spreads into the whole enzymatic sequence between PSII and PSI. Light absorption in PSI reoxidizes this sequence, probably via

P700 as first reductant, and cytochrome *f* (or plastocyanin) as the next one. Certain experiments suggest that this reaction produces a "pool" of oxidized intermediates that can survive for several seconds; the effect of PSI absorption on PSII can be, therefore, observed even after a considerable dark interval (see Chapter 13, Fig. 13.7).

The position of Q in Fig. 14.4 suggests that its redox potential should be close to 0 volt. In Chapter 14, we have identified two such compounds, cytochrome b_3 and plastoquinone (PQ). Poisoning experiments with DCMU convinced Duysens that Q is *not* plastoquinone, but precedes it in the reaction sequence. However, there are suggestions that Q is another kind of quinone. A possible relation of Q to cytochrome b_3 remains unclear.

DELAYED FLUORESCENCE

An interesting phenomenon was discovered in 1951 by Bernard Strehler and William Arnold at Oak Ridge. They found that plants and bacteria emit light quanta for quite some time (many seconds) after the exciting light had been removed. This "delayed fluorescence" has a spectrum approximately identical with that of true chlorophyll *a* fluorescence; it is thus due to delayed formation of excited Chl *a* molecules. Delayed emission (at a certain time) shows a two-light effect: light absorbed in PSI quenches emission caused by excitation of PSII. The intensity of delayed emission is very weak (it can be discovered only by means of a "quantum counter," responding to a few photons); but because of its long duration, the contribution of this emission to steady-state fluorescence may be not negligible. The molecular mechanism of delayed emission is not yet clear. (The same is generally true of "chemoluminescence"—emission of visible quanta during chemical reactions.)

The occurrence of delayed fluorescence has led Arnold to look at photosynthetic units as pseudo-crystalline bodies, in which electrons liberated by light absorption can get stuck in "traps" or irregularities of the lattice, to be released at some later time. To what extent this "solid state picture" of the photosynthetic system is significant, remains an arguable question. (It is related to the alternative: "exciton migration or electron migration" mentioned in Chapter 12.)

Chapter 16

Separation of the Two Pigment Systems

The two photochemical systems that we were led to postulate in the preceding chapters, must be closely integrated structurally and functionally with each other, and with at least three enzymatic "conveyor belts" operating in photosynthesis. Nevertheless, one can try to disintegrate photosynthetic organelles (chloroplasts) by mechanical (or chemical) means, in the hope of separating the two postulated systems, at least partially, from each other. Such fractionations have been, in fact, attempted with promising results. Their success can be judged by two criteria: *analytical*—that is, observing differences in chemical composition of the several fractions; and *functional*—by studying differences in their biochemical or photochemical activity.

Among early relevant observations is the finding that phycobilins (which form the bulk of pigment system II in red and blue-green alga) can be extracted into distilled water, leaving chlorophyll *a* in the cells.

Another early observation, by A. A Krasnovsky (in Moscow), and C. S. French (at Stanford), was that in green cells, a minor Chl component is preferentially bleached by strong red light. That a Chl component (Chl *a* 695) is preferentially damaged during ultrasonic disintegration of a suspension of *Chlorella* cells, particularly in an acid, aerobic medium, was already mentioned in Chapter 15.

Jan B. Thomas (at Utrecht, Holland) and Carl Cederstrand (in our laboratory) noted that extracting chloroplasts with aqueous methanol or acetone of different strength, leads to fractions with different absorption and fluorescence spectra, suggesting preferential extraction of one of the two pigment systems. Extracts made with 10% methanol show similarity with unextracted chloroplasts—the red absorption band is located at about 675 nm and the fluorescence yield is about 3%. As the concentration of methanol goes up from 10 to 20 and 30%, the fluorescence maxima in the extracts are shifted towards the shorter waves, and the width of the fluorescence band increases. The greatest changes are shown by the fluorescence spectra at low temperature (77°K); for example, extracts in 10% methanol strongly emit the 698 nm band (tentatively attributed to the "trap" in pigment system II; see Chapter 15); while extracts in 20% methanol emit, at 77°K, about equally intense bands at 685 nm (system II), 695 nm (system II) and 710 nm (system I). The emission spectra of the extract in 30% methanol has a main peak at 670 nm (due to dissolved Chl *a*), another at 685 nm (bulk of Chl *a* in PSII) and some emission at 695 nm (trap in PSII) and 710 nm (PSI). Photochemically, the extracts in 10–20% solvent are as active in the Hill reaction (dye reduction in light) as the chloroplasts themselves. Provisional conclusion from these observations is that pigments belonging to PSII are extracted more easily, at lower solvent concentrations, than pigments contained in PSI.

Several attempts have been made to separate the pigment systems by physical means. One of the earliest ones, by Mary Belle Allen, involved repeated freezing and grinding, followed by sonication and a so-called "density gradient centrifugation." In this way, she obtained particles enriched in chlorophyll *b* and Chl *a* 670 (that is, in PSII). More successful have been recent studies by N. K. Boardman and Jan Anderson in Australia. They disintegrated spinach leaves in a blender, filtered the juice through muslin, and precipitated the material suspended in the filtrate by centrifugation at 100 × g.[1] After resuspension of the pellet, they added a detergent, digitonin, let it act for 30 minutes at 0°C, and fractionated the suspension by increasingly strong centrifugation—from 16 minutes at 1000 × g to 60 minutes at 144,000 × g.

The two extreme fractions—the "heavy" one (HF) precipitated already at 10,000 × g and the "light" one (LF) precipitated only at 144,000 × g,

[1] One g means centrifugal force equal to gravity force on the surface of the earth.

showed significant differences in composition (see Table 16.1). For example, the concentration ratio [Chl *a*] : [Chl *b*] was 2.3 in HF, as against 5.3 in LF. (We have concluded in earlier chapters that Chl *b* is relatively more abundant in PSII.) LF was richer than HF in P700 (the "PSI trap"). The concentration of cytochrome *b* was three times higher in HF, while the concentration of cytochrome *f* was somewhat higher in LF. (In scheme 14.4, we placed cyt b_3 close to PSII and cyt *f* close to PSI.) A significant difference in the ratio [carotenol] : [carotene] also was noted—3.3 in HF and 1.9 in LF—indicating that different carotenoids may be preferentially associated with the two systems. The content of iron was 1.5 times higher in LF than in HF. (PSI is supposed to be associated with the iron-containing enzymes ferredoxin, cytochrome b_6, and cytochrome *f*; while PSII is associated with only one such enzyme, cytochrome b_3.) Finally, the manganese content of HF was 5 times that of LF. Manganese is supposed to participate in the oxygen-liberating enzymatic reaction, sensitized by PSII. All this suggested preferential accumulation of PSI in LF and of PSII in HF.

Functional tests revealed a capacity of the heavy fraction to reduce the dyes di- or trichlorophenol indophenol in light, but little capacity for reducing pyridine nucleotide ($NADP^+$) with H_2O as the electron donor; the light fraction did reduce $NADP^+$ if reduced dichlorophenol indophenol was supplied as reductant, but did not reduce the oxidized form of dichlorophenol indophenol. This difference agrees with the assumption that HF contained more PSII and LF more PSI.

The *fluorescence yield* of HF was five times that of LF—and we recall that most chlorophyll fluoresence is ascribed to PSII. We found that the fluorescence of LF, excited by polarized light, to be more strongly polarized than that of HF (5.4% versus 2.7%); and we have seen in Chapter 15 that polarization is strongest in the long-wave emission band F720, probably originating in PSI.

We—as well as Boardman and co-workers—found that at 77°K, the LF emitted more strongly at 720 nm than at 695 nm, whereas the HF emission spectrum contained a strong 695 nm band, again in agreement with preferential assignment of PSII to HF, and of PSI to LF.

All these findings—summarized in Table 16.1—suggested that Boardman and Anderson's precedure does, in fact, lead to at least partial separation of PSI and PSII. The "light" fraction may contain mostly PSI, while the "heavy" one is significantly enriched in PSII (perhaps,

TABLE 16.1 Fractionation of Chloroplast Material

Property	Light Fraction	Heavy Fraction
Chl *a*/Chl *b* ratio	5.3	2.3
Relative concentration of cytochrome *b*	1	3
Carotenol/carotene ratio	1.9	3.3
Relative Fe content	1–5	1
Relative Mn content	1	5
Capacity for reduction of TCPIP (or DCPIP)	Low	High
Capacity for reduction of $NADP^+$ at the cost of reduced DCPIP	High	Low
Percent of polarization of fluorescence excited with polarized light	5.4	2.7
Quantum yield of fluorescence	Low	High
Emission spectra at 77°K	more F720	more F696

to about 70% of the total). Recently, a great deal of other detergents have been tried; one of the more successful ones was Triton X-100, used in L. P. Vernon's laboratory at Yellow Springs, Ohio. In one case, partial separation of the pigment systems was obtained even without the use of detergents.

Jean-Marie Briantais (at Gif-Sur-Yvette) was able to obtain partial reconstruction of the properties of the original chloroplast material when the separated particles of the two types were mixed together.

Chapter 17

The Enzymatic Paths from Water to Molecular Oxygen and from Carbon Dioxide to Carbohydrates

According to Figs. 5.4 and 14.4, the main products of the two primary photochemical reactions are a reduced intermediate, XH, and an oxidized intermediate, Z.

The oxidant, Z, must be strong enough to oxidize H_2O to molecular oxygen. The reductant XH (or XH_2) must be strong enough to reduce (either alone or with the help of ATP formed as a side product of the photochemical reaction) carbon dioxide to carbohydrate.

PATH TO MOLECULAR OXYGEN

We know next to nothing about the enzymatic mechanism of oxygen liberation. (Not too much more is known about the reverse process in respiration—the uptake of molecular oxygen!) That it *does* involve enzymes can be inferred, for example, from the similarity of the saturation ratios (and the effects on them of temperature and certain poisons) of photosynthesis and of the Hill reaction (Chapters 6 and 7). These rates appear to be limited by an enzymatic reaction in the oxygen-evolving stage, common to both processses, rather than a reaction in the carbon dioxide-reducing stage, which is nonoperative in the Hill reaction.

The enzymatic system involved in the liberation of oxygen seems to require *manganese*—probably, as a heavy metal component of some enzyme. From a variety of experiments, it is now clear that manganous and chloride ions are associated with the operation of light reaction II. For example (see Chapter 16), manganese is present in higher concentration in the "heavy" fraction of chloroplast particles, which perform preferentially light reaction II, than in the "light" fraction, which sensitizes preferentially light reaction I. Manganese deficiency was found not to affect the chloroplast-sensitized photoreduction of pyridine nucleotide ($NADP^+$) by reduced dichlorophenol indophenol—a reaction apparently involving only PSI; while it did stop the complete Hill reaction (which requires the operation of both PSI and PSII). The high oxidation-reduction potential of the couple Mn^{3+}/Mn^{2+} (E_0' about 1.5 volts!) makes it a plausible speculation to suggest that Mn^{2+} is photooxidized to Mn^{3+} in light reaction II, and Mn^{3+} then oxidizes H_2O by a dark reaction ($2Mn^{3+} + H_2O \rightarrow \frac{1}{2}O_2 + 2Mn^{2+} + 2H^+$). Of course, Mn^{3+}, complexed with proteins in an enzyme, is likely to have a redox potential considerably less positive than that of the free Mn^{3+}/Mn^{2+} couple—but it may still be sufficient to oxidize water to oxygen (requiring $E_0' > +0.75$ volt).

Recent kinetic studies in P. Joliot's laboratory in Paris and in J. Rosenberg's laboratory in Pittsburgh have provided some information regarding details of the oxygen evolution process. They showed that the formation of *one* oxygen molecule requires the accumulation of *four* precursor molecules, formed by four light reactions.

THE JUNCTION OF THE PHOTOCHEMICAL PROCESS WITH THE ENZYMATIC SEQUENCE OF THE CO_2 REDUCTION

In schemes 5.4 and 14.4, we designated the compound located in the junction of the vertical arrow with the upper horizontal arrow as "X."

This compound is supposed to be reduced, in photochemical reaction I, to XH (or XH_2); the latter must bring about the reduction of CO_2 by an enzymatic reaction sequence.

We do not know the chemical identity of X. Since the presence of $NADP^+$ in green cells is well established, and chloroplasts are known to be able to use this compound as a "Hill oxidant," it had been suggested that X is identical with $NADP^+$ (E_0' is —0.35 volt for the couple $NADP^+$/NADPH). However, more recently, it has been made likely that (at least) two NADPH-precursors—an iron-containing protein, "ferredoxin," and an enzyme, "ferredoxin-NADP-reductase" (mediating the reduction of $NADP^+$ by reduced ferredoxin), precede $NADP^+$ in the photosynthetic reaction sequence. Is ferredoxin ($E_0' =$ —0.4 volt) then the primary photochemical oxidant, X? Bessel Kok (at the RIAS laboratory in Baltimore), found, that illuminated chloroplasts can reduce dyes with E_0'-values down to —0.6 volt; X may be a compound with a similarly low potential. Ferredoxin, Fd, would then be an intermediate between X and NADPH:

$$\frac{\text{X}}{\text{XH}} \rightarrow \quad \frac{\text{Fd}}{\text{reduced Fd}} \rightarrow \begin{matrix}\text{Fd-NADP}\\ \text{reductase}\end{matrix} \rightarrow \frac{\text{NADP}^+}{\text{NADPH}} \qquad (17.1)$$
$$E_0' = -0.6\ \text{volt} \quad -0.4\ \text{volt} \qquad\qquad -0.35\ \text{volt}$$

The reduction of $NADP^+$ involves, at pH 7, the transfer of one H-atom and one electron ($\bar{e}$):

$$\text{NADP}^+ + \text{H} + \text{e}^- \rightarrow \text{NADPH}$$

or

$$\text{RN}^+\!\begin{matrix}\diagup\text{CH=CH}\diagdown\\ \diagdown\!\!\diagdown\text{CH—C}\diagup\!\!\diagup\\ \qquad\diagdown\text{ONH}_2\end{matrix}\!\text{CH} + \text{H} + \text{e}^- \rightarrow \text{RN}\!\begin{matrix}\diagup\text{CH=CH}\diagdown\\ \diagdown\text{CH}_2\text{—C}\diagup\!\!\diagup\\ \qquad\diagdown\text{ONH}_2\end{matrix}\!\text{CH} \qquad (17.2)$$

where R is an organic residue containing 3 phosphate groups, 2 ribose groups (ribose being a five-carbon sugar) and a purine base called adenine (see Chapter 18).

Perhaps, in the "downhill" redox reaction (17.1), enough energy becomes available to produce, without extra expenditure of light energy, a third "bonus" ATP (in addition to the two provided by four electrons sliding down the intermediate enzyme reaction sequence, from Cyt *b* to Cyt *f*, that is, from $E_0' = 0.0$ to E_0' $+0.4$ volts). This would be a welcome possibility from the point of view of Calvin's scheme (to be described later in this chapter), since the latter calls for three ATP molecules to be invested in the reduction of each CO_2 molecule (see Chapter 18).

A theoretical alternative is direct use of XH_2 for reduction of a carbonyl group in phosphoglyceric acid (PGA) (avoiding altogether Fd and $NADP^+$ as intermediates!); this would eliminate the need for the two ATP molecules required for this key step of photosynthesis. (Since $NADP^+$ has an E_0' of only $-.035$ volt, while E_0' of the carboxyl/carbonyl couple is about -0.5 volt, the reduction of carboxyl to carbonyl by NADPH cannot take place without the assistance of an ATP molecule!) This alternative deserves to be kept in mind.

In certain blue-green algae, ferredoxin is replaced by a flavin-containing protein "flavodoxin." (Flavins are the well-known water-soluble yellow redox intermediates involved in the respirations of plants and animals.)

THE PATH OF CARBON IN PHOTOSYNTHESIS

The conversion of CO_2 to carbohydrate with the help of a reducing agent (such as NADPH) provided by the photochemical reaction, is the most thoroughly analyzed stage in photosynthesis. This success had been due to the availability of a convenient tracer—a radioactive carbon isotope. It is well known that such tracers can be used to follow the pathway of various elements in chemical reactions, by observing the appearance of their characteristic radiations in different reaction intermediates and products. The application of this method to photosynthesis has been one of its most spectacular successes, rewarded in 1961 by the Nobel Prize given to Melvin Calvin of Berkeley, California. Together with A. A. Benson, J. Bassham and other co-workers, he has achieved most important progress in this field.

The ordinary, nonradioactive, carbon isotope is ^{12}C (the superscript referring to the atomic mass). The first radioactive carbon isotope used as tracer was ^{11}C; however, its application had only limited success, because of its very rapid decay (^{11}C has a half-time of 20.5 sec). The discovery of the long-lived ^{14}C (half-life, 5720 years) by Martin Kamen and Sam Ruben at Berkeley (California) just before World War II, made it possible to follow, at leisure as it were, the passage of the carbon tracer from a "tagged" substrate (such as carbon dioxide or bicarbonate) into one organic compound after another. The ^{11}C isotope was given up as an analytical tool, and ^{14}C became *the* all-important carbon tracer. We will use for it the simpler symbol C*, with the asterisk indicating its radioactive nature.

The first applications of C* to photosynthesis led to confusing results, until two facts were established; one, that photosynthesis in live cells is rapidly followed by secondary transformations, bringing C* into a variety of metabolic products; and two, that CO_2 is not such a biochemically inert compound as it had been previously thought to be. Because of the latter fact, the tracer, supplied as C^*O_2 or $HC^*O_3^-$, can enter certain metabolic products by dark enzymatic reactions. In particular, the carboxyl groups of certain organic acids, such as oxalacetic acid and pyruvic acid (which are important intermediates in respiration), undergo relatively rapid isotopic exchange with external carbon dioxide:

$$RCOOH + CO^*_2 \rightleftharpoons RC^*OOH + CO_2$$

(where the radical R is $HOOC{-}CH_2{-}C{=}O$ in oxalacetic acid, and $CH_3{-}C{=}O$ in pyruvic acid). Exchange reactions of this type cause a "nonreductive" incorporation of radioactive carbon into carboxylic acid, and thence into other metabolic compounds, such as amino acids.

Rapid progress first became possible when Calvin and co-workers recognized that in order to separate the fast photochemical incorporation of C* into organic material from slower incorporation by dark reactions, one must work very rapidly. They proceeded to expose already actively photosynthesizing plants (to avoid induction delays) to $HC^*O_3^-$ or C^*O_2 for seconds or minutes rather than for hours (as it was done before). Such fast experiments were made possible by two new, highly sensitive analytical techniques, paper chromatography and autoradiography.

In these experiments, a suspension of algae (or another plant material) is exposed to light in "ordinary," nonradioactive carbon dioxide or carbonate until a steady state of photosynthesis is established. Radioactive carbon is then supplied by sudden injection of C*-containing bicarbonate into the suspension medium (or of C^*O_2 into the circulating gas). Illumination is continued for a few seconds (or minutes); then, plant cells are rapidly killed—usually by dropping them into boiling alcohol. The cell material is extracted with various solvents, and the extract fractionated by so-called two-dimensional paper chromatography.

In this procedure, excess solvent is removed by evaporation, and the concentrated extract applied near one of the corners of a sheet of filter paper. Then, a suitable solvent is allowed to mount through the paper, driven upward by capillarity; different components of the solution are deposited at different heights, depending upon their properties, thus forming a one-dimensional sequence along one edge of the sheet. The paper is then dried, and the edge along which the compounds had been deposited is dipped into another solvent. The compounds are now carried up to different heights at right angle to the direction of their original diffusion, thus forming a two-dimensional "chromatogram."

This procedure can lead to a very effective fractionation. With one basic and one acidic solvent, neutral components (such as sugars or esters) may be deposited near the original corner of the sheet, while basic and acidic components deviate in opposite directions from the diagonal. How far the different components diffuse depends on their adsorbability on paper and their solubility in the solvents used. By using mixtures of known chemical compounds, one can prepare an empirical "chromatographic map," showing where these compounds will be found after a certain period of diffusion (at a certain temperature, on a certain kind of filter paper).

Because of the radioactivity of the components that form the subject of this research, their position on the paper chromatogram can be identified by "autoradiography," even if their amounts are so small as to be invisible to the eye. For this, the filter paper is pressed against a sheet of X-ray film, and the latter is developed after proper exposure. The radiation (β^- particles), emanating from the radioactive carbon present in certain spots on filter paper, blackens the opposite spots on the X-ray film (see Fig. 17.1).

Various improvements of this wonderfully sensitive and powerful analytical method have been introduced. The amount of C* in the various spots can be estimated from the intensity of radiation emitted by them.

The spots can be treated by analytical reagents; they can be cut out, dissolved, and rechromatographed, or investigated by chemical methods.

The First Stable Product of CO_2 Fixation: Phosphoglyceric Acid[1]

The above-described combination of two ingenious techniques proved a unique boon to the study of the mechanism of carbon dioxide reduction

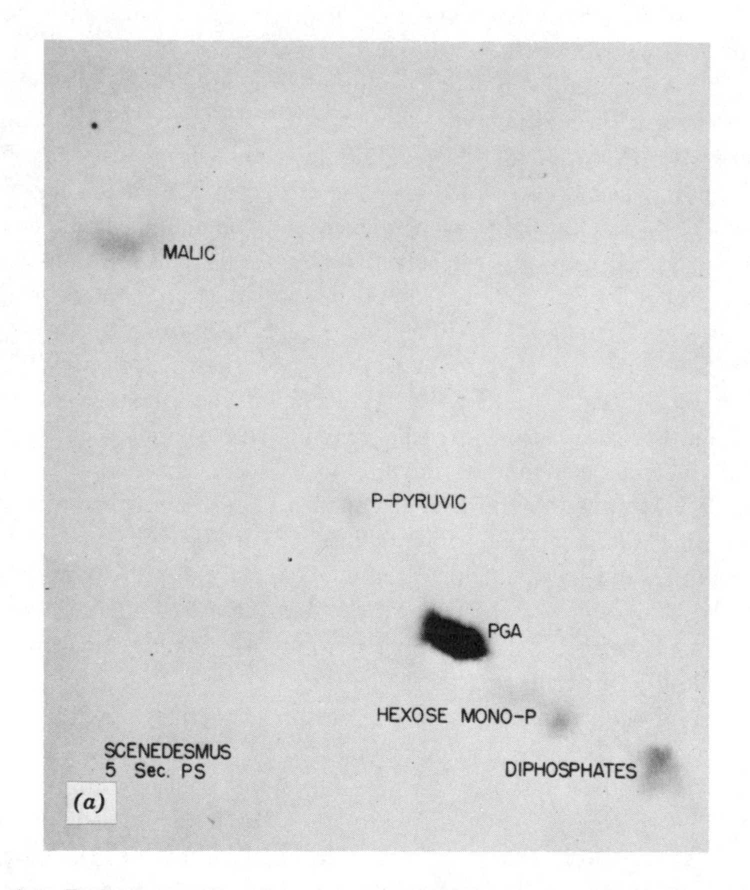

FIG. 17.1 Radiograms of products photosynthesized by Scenedesmus in CO_2 (see p. 227). (*a*) in 5 sec; (*b*) in 15 sec; (*c*) in 60 sec. The "neutral" products, such as sugar phosphates (designated by the letter *P*) remain nearest to the corner, while basic compounds are carried to the left and acidic ones to the right of the diagonal. (M. Calvin and co-workers, 1951.)

[1] The rest of this chapter requires some familiarity with organic chemistry and biochemistry.

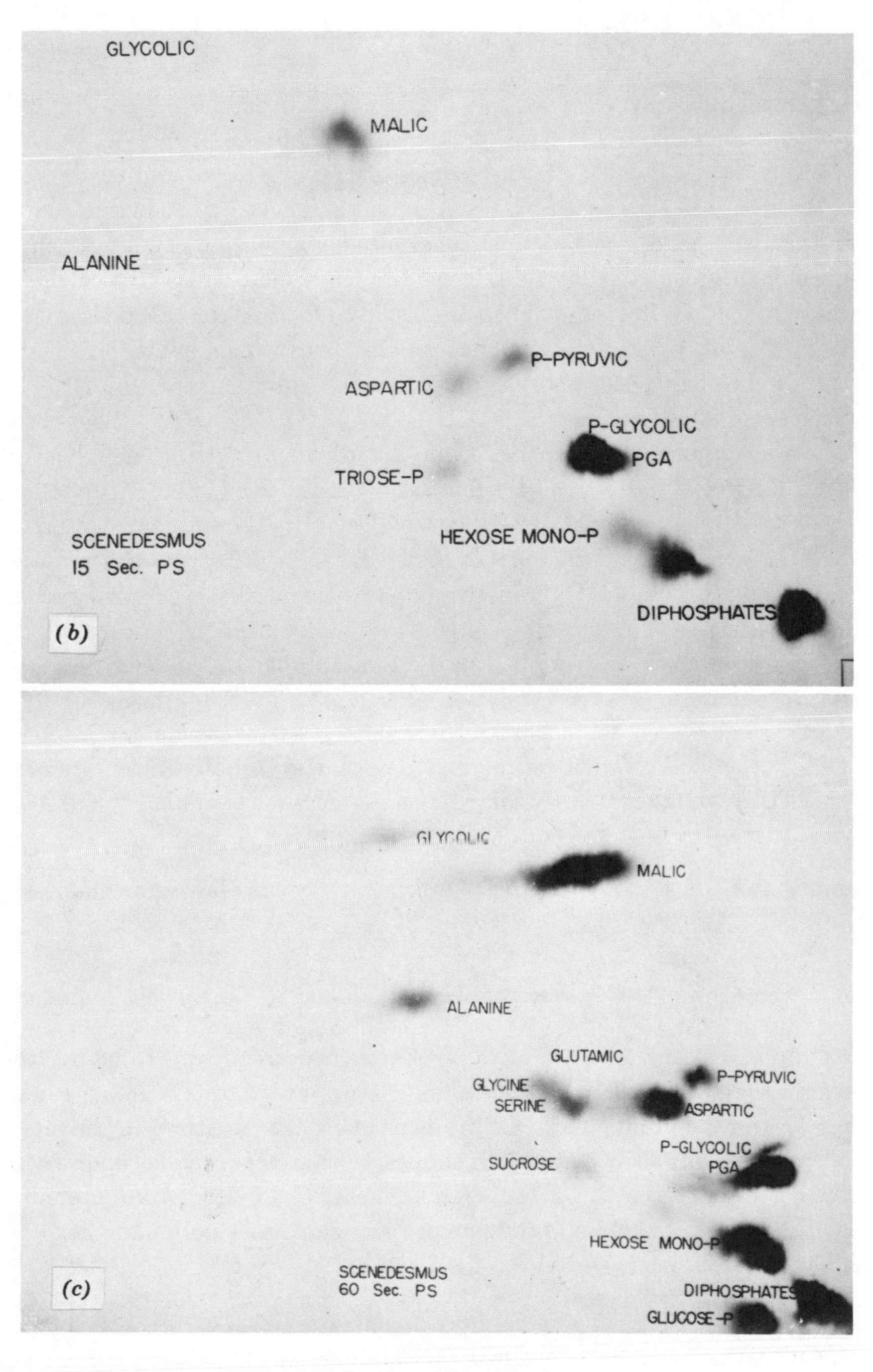

FIG. 17.1 (Continued)

in photosynthesis (which was, at that time, still dominated by the formaldehyde-formose hypothesis mentioned in Chapter 6!).

The main results of the initial studies of Melvin Calvin and co-workers at Berkeley are illustrated by Figs. 17.1*a,b* and *c,* which show the "chromatographic map" of the C*-tagged products of photosynthesis in a unicellular green alga after 5 seconds, 15 seconds, and 60 seconds of exposure to C^*O_2 in strong light, respectively. One sees that after 5 seconds, over 90% of C* is found in one single spot, which has been identified as belonging to phosphoglyceric acid (abbreviated PGA). The formula of this compound is $CH_2O(P) \cdot CHOH \cdot COOH$.[2] (The name "glyceric" indicates derivation from the three-carbon alcohol glycerol, $CH_2OH \cdot CHOH \cdot CH_2OH$.) Further analysis showed that after short exposure of cells to light, all C* is contained in the carboxyl group of PGA molecule; $CH_2O(P) \cdot CHOH \cdot C^*OOH$. Only after several minutes of illumination does C* gradually appear also in the other two carbon positions.

After 60 seconds, we see in Fig. 17.1*c* a number of radioactive spots, corresponding to a variety of tagged compounds. Some are acid (like PGA), some basic, some neutral. The latter are of particular interest, because the final products of photosynthesis, the carbohydrates, are neutral. The carbohydrates (sugars) have the general formula $(C_m(H_2O)_n)$, and contain either an aldehyde group $-C\overset{\diagup H}{\underset{\diagdown\!\diagdown O}{}}$ if they are "aldoses"; or a keto group $\diagdown\!\!\!\diagup C{=}O$ if they are "ketoses." The earliest and most heavily labeled neutral products of photosynthesis proved to be not the hexose sugars themselves (the ending "ose" designates a simple sugar, $m = n$ and the prefix "hex" the presence of six carbon atoms in the molecule $n = 6$), but their *phosphates,* that is, hexose esters of phosphoric acid (see bottom right corner of Fig. 17.1*c*). These hexose derivatives probably are precursors of the final products of photosynthesis–polymeric hexoses, such as starch.

[2] (P) stands for the univalent phosphoric acid residue H_2PO_3. The acid itself is (P)OH; the notation Pi is often used instead to denote "inorganic" phosphate (as contrasted to "high energy phosphates," such as ATP, see Chapter 18).

Among hexoses, the most important are glucose (an aldose) and fructose (a ketose), shown in Eq. 17.3.[3]

$$
\begin{array}{c}
\begin{array}{rcl}
\text{H}- & \text{C} & =\text{O} \\
 & | & \\
\text{H}- & \text{C} & -\text{OH} \\
 & | & \\
\text{HO}- & \text{C} & -\text{H} \\
 & | & \\
\text{H}- & \text{C} & -\text{OH} \\
 & | & \\
\text{H}- & \text{C} & -\text{OH} \\
 & | & \\
 & \text{CH}_2\text{OH} & \\
\end{array} \\
\text{Glucose} \\
\text{(aldose)}
\end{array}
\qquad
\begin{array}{c}
\begin{array}{rcl}
 & \text{CH}_2\text{OH} & \\
 & | & \\
 & \text{C} & =\text{O} \\
 & | & \\
\text{HO}- & \text{C} & -\text{H} \\
 & | & \\
\text{H}- & \text{C} & -\text{OH} \\
 & | & \\
\text{H}- & \text{C} & -\text{OH} \\
 & | & \\
 & \text{CH}_2\text{OH} & \\
\end{array} \\
\text{Fructose} \\
\text{(ketose)}
\end{array}
\qquad (17.3)
$$

Cane sugar (sucrose), $C_{12}H_{22}O_{11}$, is the product of condensation of one molecule of glucose with one molecule of fructose (with loss of a water molecule); starch is a high-polymeric form of glucose.

One sugar phosphate spot on the chromatogram of early products of photosynthesis proved to be peculiarly important. Identification of this product as a *pentose phosphate* (that is, a phosphate of a sugar containing five rather than six carbon atoms)—more specifically, as *ribulose diphosphate* (for short, RDP), by A. A. Benson and co-workers in 1951, was the second fundamental discovery resulting from the application of radiocarbon tracer to photosynthesis (after the identification of PGA). Ribulose diphosphate turned out to be the main or only carbon dioxide "acceptor" in photosynthesis.

The Primary CO_2 Acceptor: Ribulose Diphosphate

It is a priori unlikely that carbon dioxide could undergo reduction as such (or as a carbonate ion); rather, it was long surmised that CO_2 is first bound in photosynthesis by a nonphotochemical reaction to some organic "acceptor."

After the identification of PGA as the most important early-labeled intermediate of photosynthesis, the nature of the CO_2 "acceptor" suggested itself by the following consideration: Since at first, PGA carries

[3] The OH and H groups are placed to suggest the spatial arrangement of the substituents in the specific forms of the sugars found in nature ("dextro" forms). Their mirror image stereomers ("levo" forms) do not occur in nature.

the "label" entirely in its carboxyl group, it must have been formed by *carboxylation,* according to the equation: $RH + C^*O_2 \rightarrow RC^*OOH$ (a reaction in which CO_2 inserts itself between an organic radical R, and an H-atom). It seems that to produce PGA by such a reaction, the acceptor should be *glycol phosphate,* $CH_2O(P) \cdot CH_2OH$ (glycol is the two-carbon di-alcohol, $CH_2OH—CH_2OH$). This presumption could not be confirmed; but the riddle solved itself by identification of ribulose diphosphate (RDP) as an early-tagged product of photosynthesis. When RDP (a five-carbon compound) adds a CO_2 molecule, an acid containing six carbon atoms is formed ($C_5 + C_1 = C_6$). If a molecule of this acid breaks into two parts, each containing three carbon atoms, both can be PGA molecules (as indicated in Eq. 17.4)!

$$
\begin{array}{c}
H_2CO(P) \\ | \\ C{=}O \\ | \\ HC{-}OH \\ | \\ HC{-}OH \\ | \\ H_2CO(P) \\ \\ \boxed{\text{RDP}}
\end{array}
\quad
\xrightarrow[\text{(carboxylation)}]{+C^*O_2}
\quad
\begin{array}{c}
H_2C{-}O\ (P) \\ | \\ HOOC^*{-}C{-}OH \\ | \\ C{=}O \\ | \\ HC{-}OH \\ | \\ H_2CO(P) \\ \\ \boxed{\begin{array}{c}C_6 \text{ addition} \\ \text{compound}\end{array}}
\end{array}
\quad
\xrightarrow[\substack{\text{(hydrolysis} \\ \text{and} \\ \text{dismutation)}}]{+H_2O}
\quad
\begin{array}{c}
H_2CO(P) \\ | \\ HC{-}OH \\ | \\ C^*OOH \\ + \\ COOH \\ | \\ HC{-}OH \\ | \\ H_2CO(P) \\ \\ \boxed{\text{2 PGA}}
\end{array}
\qquad (17.4)
$$

Reaction 17.4 involves, in addition to *carboxylation* (addition of CO_2) and splitting of the six-carbon chain in two by hydrolysis, also a *"dismutation"*—oxidation of one carbonyl group to carboxyl, and reduction of another one to alcohol. (Otherwise, a single CO_2 could not produce *two* carboxyl groups!) The enzyme that brings about this combined reaction is called a *carboxy-dismutase.*

If RDP is the primary acceptor of CO_2, and PGA is formed as the product of this carboxylation, a sudden depletion of C^*O_2 in a steadily photosynthesizing system should result in an immediate drop of the PGA concentration, and corresponding rise of that of RDP. This was confirmed in Calvin's laboratory (see Fig. 17.2*a*).

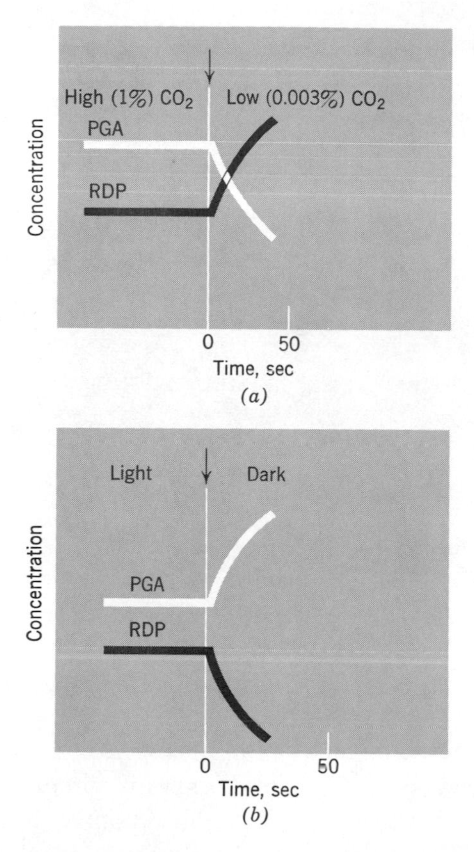

FIG. 17.2 Diagram showing the effect of lowering the CO_2 level (*a*) and turning off the light (*b*) on the concentration of phosphoglyceric acid (PGA) and ribulose diphosphate (RDP). (M. Calvin and co-workers, 1957.)

The reduction of PGA by a reductant supplied by the photochemical primary process leads in part to the formation of sugar, and in part to regeneration of the CO_2 acceptor used up in the reaction. More specifically, five sixths of reduced PGA must be reconverted to RDP and one sixth—corresponding to the one molecule CO_2 added in carboxylation (17.4)—can be converted to the final product, hexose. The resulting reaction sequence is known as the "Calvin cycle." It is represented in Figs. 17.3 and 17.4.

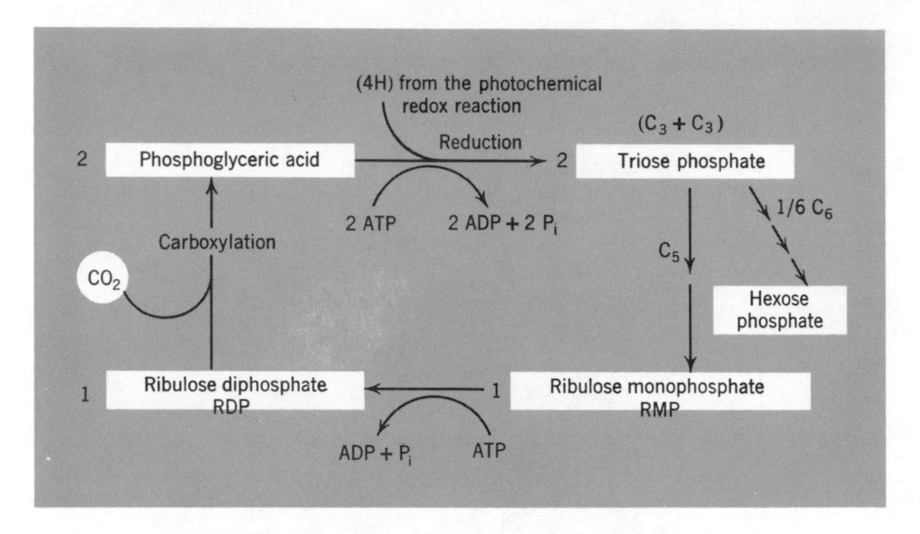

FIG. 17.3 A simplified version of the path of carbon in photosynthesis (Pi = (P)OH = inorganic phosphate). (After M. Calvin and co-workers.)

The Reduction of Phosphoglyceric Acid

Carboxylation must be followed in photosynthesis by *reduction* of the carboxylated acceptor by a photochemically produced reductant. As mentioned before, most workers are inclined to consider reduced pyridine nucleotide, NADPH, as the actual reductant. As pointed out before, reduced pyridine nucleotide does not have enough reducing power (its potential, $E_0' = -0.35$, is not sufficiently negative), to reduce a carboxyl group to a carbonyl group (the E_0' of the couple carboxyl-carbonyl is about -0.5 volt). It has been suggested that this reduction becomes possible if it is coupled with the energy-releasing hydrolysis of a molecule of a "high energy phosphate": $\text{ATP} \xrightarrow{+\text{H}_2\text{O}} \text{ADP} + \text{P(OH)}$ (see Chapter 18).

Calvin and co-workers suggested that phosphoglyceric acid (PGA), formed in reaction 17.1, is phosphorylated, in the next reaction step, to diphosphoglyceric acid (DPGA) (see Eq. 17.5). The latter is a "high energy" phosphate, because the second (P) radical is attached to the carboxyl group (see Chapter 18). This reaction is catalyzed by an enzyme of the type called "kinases." DPGA can be reduced by NADPH (see Eq. 17.6), if its high energy phosphate group is simultaneously split

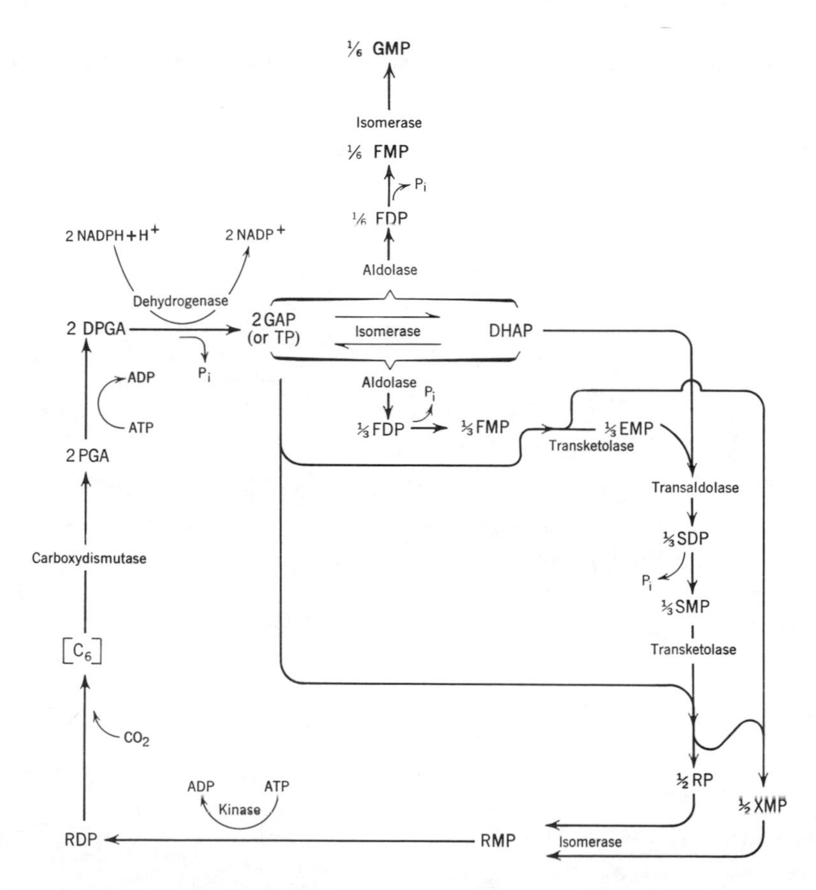

FIG. 17.4 The Calvin-Benson-Bassham cycle. (Pi = (P)OH = inorganic phosphate).

C_3 Compounds

PGA: 3-phosphoglyceric acid

DPGA: 1,3-diphosphoglyceric acid

GAP (or TP): glyceraldehyde-3-phosphate mono phosphate

DHAP: dihydroxyacetone phosphate

C_4 Compounds

EMP: erythrose-4-mono phosphate

C_5 Compounds

RP: ribose-5-mono phosphate

RMP: ribulose-5-mono phosphate

RDP: ribulose-1,5-diphosphate

XMP: xylulose-5-mono phosphate

C_6 Compounds

FMP: fructose-6-mono phosphate

FDP: fructose-1,6-diphosphate

GMP: glucose-6-phosphate

C_7 Compounds

SMP: sedoheptulose-7-mono phosphate

SDP: sedoheptulose-1,7-diphosphate

off as a "low energy" phosphate, (P)OH. The product of this reduction is a triose—a sugar with three carbon atoms (or rather, a triose phosphate, TP).

$$\begin{array}{l} CH_2O(P) \\ \quad | \\ HC{-}OH \\ \quad | \\ \;C{-}OH \\ \quad \| \\ \quad O \end{array} + ATP \xrightarrow{\text{(Kinase)}} \begin{array}{l} CH_2O(P) \\ \quad | \\ HC{-}OH \\ \quad | \\ \;C{-}O(P) \\ \quad \| \\ \quad O \end{array} + ADP$$

$$\boxed{\text{PGA}} \qquad\qquad \boxed{\text{DPGA}} \qquad\qquad (17.5)$$

Reaction 17.6 requires an enzyme called triose phosphate dehydrogenase. (Note that the names of enzymes usually reveal the substrates they act on, and the kind of ractions they catalyze.)

$$\begin{array}{l} CH_2O(P) \\ \quad | \\ HC{-}OH \\ \quad | \\ \;CO(P) \\ \quad \| \\ \quad O \end{array} + NADPH + H^+ \xrightarrow{\text{(Dehydro-genase)}} \begin{array}{l} CH_2O(P) \\ \quad | \\ HC{-}OH \\ \quad | \\ \;CH \\ \quad \| \\ \quad O \end{array} + (P)OH + NADP^+$$

$$\boxed{\text{DPGA}} \qquad\qquad \boxed{\text{TP}} \qquad\qquad (17.6)$$

Reaction 17.6 utilizes the reducing power of NADPH, while reaction 17.5 makes use of the high energy content of ATP—both acquired in the light reactions of photosynthesis.

If these reactions occur as suggested, turning off the light illuminating a steadily photosynthesizing cell suspension should lead to an immediate transitory increase in the concentration of PGA, and a decrease in that of RDP, because the reactions 17.4 will continue for a while, while reactions 17.5 and 17.6 must stop when no more ATP and NADPH are supplied by light. This expectation was confirmed (see Fig. 17.2*b*).

As mentioned above, one sixth of triose phosphate formed by reduction of PGA must be converted, by a series of enzymatic reactions, into a hexose, which is the final product of photosynthesis; while five sixths must be tranformed back into ribulose phosphate, to close the cycle (Fig. 17.3). Each time the cycle in Fig. 17.3 revolves once, one sixth of a hexose molecule is formed, and one molecule CO_2 and four hydrogen atoms are consumed

$$CO_2 + (4H) \rightarrow (CH_2O) + H_2O$$

After the cycle had rotated six times, one hexose molecule is formed, and six CO_2 molecules are used up.

The Transformation of Carbohydrates

It was stated above that the triose, $C_3H_6O_3$ (glyceraldehyde) is a carbohydrate: $(CH_2O)_3$. Its transformation into other carbohydrates, such as hexoses (glucose or fructose), or into a pentose (such as ribulose), which all have the same reduction level, $R = 1$ (see Chapter 5), can occur by enzymatic conversions with no additional energy requirement. However, experience shows that one ATP molecule is needed for the conversion of ribulose monophosphate into ribulose diphosphate.

The specific mechanism by which triose phosphate (TP or GAP, glyceraldehyde phosphate), is converted in photosynthesis into hexose phosphates, has been established by tracer studies. Glyceraldehyde phosphate (GAP) first undergoes partial isomerization (with the help of an enzyme called isomerase) into dihydroxyacetone phosphate (DHAP) as in Eq. 17.7. The latter is another three-carbon sugar phosphate, which contains a ketone (C=O) group instead of the aldehyde (CHO) group present in glyceraldehyde. An equilibrium is established with 60% glyceraldehyde and about 40% DHAP.

$$\begin{array}{ccc} CH_2O(P) & & CH_2O(P) \\ | & & | \\ HC\text{—}OH & \underset{\longleftarrow}{\xrightarrow{\text{isomerase}}} & C\text{=}O \\ | & & | \\ O\text{=}C\text{—}H & & CH_2OH \\ \boxed{\begin{array}{c}\text{GAP}\\ 60\%\end{array}} & & \boxed{\begin{array}{c}\text{DHAP}\\ 40\%\end{array}} \end{array} \tag{17.7}$$

One molecule of GAP and one molecule of DHAP combine (under the action of an enzyme called "aldolase") to form a molecule of fructose diphosphate (FDP).

$$
\begin{array}{ccccc}
 & & & & CH_2O(P) \\
 & & & & | \\
CH_2O(P) & & CH_2O(P) & & C{=}O \\
| & & | & & | \\
HC{-}OH & + & C{=}O & \xrightarrow{\text{aldolase}} & HO{-}C{-}H \\
| & & | & & | \\
O{=}C{-}H & & CH_2OH & & H{-}C{-}OH \\
 & & & & | \\
 & & & & H{-}C{-}OH \\
 & & & & | \\
 & & & & CH_2O(P) \\
\boxed{\text{GAP}} & & \boxed{\text{DHAP}} & & \boxed{\text{FDP}}
\end{array}
\qquad (17.8)
$$

The so-formed fructose diphosphate (FDP) loses a phosphate group (by action of the enzyme *phosphatase*), forming fructose monophosphate (FMP). With the help of an *isomerase*, the latter isomerizes partially to glucose monophosphate. Once glucose and fructose monophosphates are available, higher-molecular carbohydrates, sucrose or starch, can be formed. We do not need to enter here into the details of this synthesis. The process leading from ribulose diphosphate to glucose monophosphate is indicated in Fig. 17.4 (which is an elaboration of Fig. 17.3).

Regeneration of Ribulose Diphosphate

The conversion of one sixth of the triose formed in photosynthesis to hexose was described previously as a "straighforward" process (17.9*a*); but regeneration of a pentose, which must use up the remaining five sixths (17.9*b*), is much more involved. The net results are:

$$2C_3 \rightarrow C_6 \qquad (17.9a)$$

$$10C_3 \rightarrow 6C_5 \qquad (17.9b)$$

To carry out the conversion (17.9*b*) in steps, each involving one or two sugar molecules, required from nature real ingenuity! Biochemists working with C*-indicator, were able to identify these steps. The conclusions are illustrated in Scheme 17.10 (Fig. 17.5). A key role is played here by vitamin B_1 (thiamine), which can transfer C_2 groups from one sugar to another. It is used in two ways: it first takes a C_2 group from a C_6-sugar (formed according to 17.9*a*), forming a C_4 sugar (erythrose); and then a second one from a C_7 sugar (sedoheptulose),

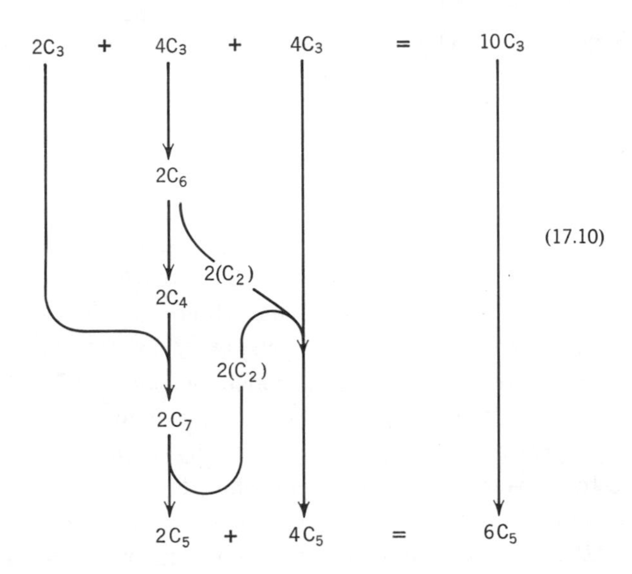

FIG. 17.5 The interconversions of carbon compounds; the subscript refers to the number of carbon atoms in particular molecules (see text).

formed by combining erythrose with a triose (C_3). This gives one pentose molecule; two more are formed by addition of the two thiamine-carried C_2-groups to triose molecules. Some additional details of scheme 17.10 are shown in Fig. 17.4. The transformation of C_3-chains into C_5-chains is shown there to include "transketolizations" and "transaldolizations," as well as a phosphorylation, producing as intermediates, in addition to erythrose (a C_4 sugar) and sedoheptulose (a C_7 sugar), also xylulose (a C_5 sugar, isomeric with ribulose). Scheme 17.10 (Fig. 17.5) describes, however, the essence of this complicated series of transformations—conversion of ten C_3-sugar molecules into six C_5-sugar molecules, by reactions involving only pairs of molecules, and a catalyst—thiamine—busily carrying C_2 groups between them.

The Calvin Cycle: Concluding Remarks

Figure 17.3 is a condensed version of the Calvin cycle—an elaboration of the upper arrow in Figs. 5.4 and 14.4; each arrow in it represents a sequence of several reactions. The three major reactions are:

1. *Carboxylation* of RDP (that is, addition of CO_2 to the 5-carbon keto sugar phosphate, RDP) and formation of PGA, a 3-carbon acid;
2. *Reduction* of PGA to a 3-carbon aldose—triose phosphate (TP)—and its polymerization to C_6-sugars;
3. *Regeneration* of RDP from TP.

As mentioned earlier, the reductant generated by the light reactions (usually presumed to be NADPH) is utilized for the reduction of PGA (reaction 2 above), while ATP (adenosine triphosphate), another product of the light reaction, is utilized at two sites: in the phosphorylation of the five-carbon keto sugar ribulose monophosphate, RMP, to diphosphate, RDP; and in the phosphorylation of phosphoglyceric acid, PGA, to diphosphoglyceric acid, DPGA, prior to its reduction by NADPH.

As suggested earlier, this second phosphorylation could be dispensed with if XH ($E_0' \simeq -0.6$ volt) could be used directly for the reduction of PGA, without some reducing power being first lost by electron transfer to $NADP^+$ ($E_0' = -0.35$ volt).

Despite the many remaining uncertainties (see below), the discoveries of Calvin, Benson, Bassham, and co-workers, particularly the identification of the two key compounds, PGA and RDP, stand as the most spectacular achievements in the study of metabolic reaction sequences by means of radioactive carbon tracer.

Alternate Pathways

Several unsolved, or only tentatively solved questions remain. Some early tagged products are alien to the cycle in Figs. 17.3 and 17.4. Among them are *carboxylic acids* (particularly malic acid, succinic acid, and glycolic acid) and *amino acids* (alanine, aspartic acid, serine, and glycine). Their appearance suggests that the cycle is complicated by various side reactions which become significant already in the first minutes of illumination. In many experiences, *malic acid* has been found heavily labeled after brief photosynthesis. Malic acid is known to be related to one of the central compounds of the Krebs cycle, pyruvic acid, $CH_3COCOOH$, by carboxylation and reduction:

$$\underset{\boxed{\text{pyruvic acid}}}{CO_2 + CH_3COCOOH} \xrightarrow{+2H} \underset{\boxed{\text{malic acid}}}{HOOCCH_2CH(OH)COOH} \qquad (17.11)$$

(Krebs cycle is the cycle involved in the oxidation of pyruvic acid to CO_2 and H_2O during respiration.) Are we to assume that reaction 17.11 provides a second "port of entry" of CO_2 into photosynthesis? Or is malic acid merely a rapidly formed by-product of the transformation of PGA?

M. D. Hatch and C. R. Slack (1966, 1968) suggested an alternate pathway of C^*O_2 fixation in corn leaves and several other plants. In this pathway, phosphoenolpyruvic acid is carboxylated as follows:

$$\begin{array}{l} CH_2 \\ \| \\ CO(P) \\ | \\ COOH \end{array} + C^*O_2 \xrightarrow[\text{Phosphoenol pyruvate carboxylase}]{} \begin{array}{l} C^*OOH \\ | \\ CH_2 \\ | \\ C{=}O \\ | \\ COOH \end{array} + P(OH) \qquad (17.12)$$

Phosphoenol pyruvic acid — Oxalacetic acid

(Malic acid is formed as a side product from oxalacetic acid by NADH; malic acid dehydrogenase is the catalyst for this reaction.) Thus, RDP is not the direct acceptor of CO_2 in this pathway. However it is suggested that RDP and the 4-carbon acid formed above (see Eq. 17.12) react together to form PGA (that enters the Calvin cycle) and pyruvic acid; the latter is then phosphorylated to phosphoenol pyruvic acid, which continues the cycle (see Equations 17.13 and 17.14).

$$\text{4-Carbon acid} + \text{RDP} \rightarrow \text{2PGA} + \text{Pyruvic acid} \qquad (17.13)$$

$$\text{Pyruvic acid} + \text{ATP} + \text{P(OH)} \xrightarrow[\text{Pyruvate synthetase}]{} \text{Phosphoenol pyruvic acid} + \text{AMP} + \text{P(P)OH} \qquad (17.14)$$

All the enzymes and the intermediates suggested by Equations 17.12–17.14 have been found in large enough quantities in corn and some, but not all, other plants studied. The author's surmised that in the latter, the cycle operates in Calvin's original form.

Formation of tagged glycolic acid ($CH_2OH \cdot C^*OOH$) in light has been noted in several laboratories. Under certain conditions (low CO_2 pressure, high concentration of O_2), it is much enhanced. The mechanism of this glycolate formation is, as yet, not properly understood.

It has been long suspected that direct photosynthesis of compounds other than carbohydrates can take place in photosynthetic systems. It has been shown, for example, that under certain conditions, 30% or more of the photosynthate in the green alga *Chlorella* appears in the form of amino acids. Under other conditions, when algal cells are exposed to C^*O_2 for only 1–2 minutes, as much as 30% of the fixed carbon is found in lipidlike substances.

One likes the simplicity of Calvin's cycle—a single carboxylation and a single reduction!—but nature is not under obligation to be simple.

Chapter 18

Photophosphorylation

HIGH ENERGY PHOSPHATE

High energy phosphate is a "portable battery charger," an almost omnipresent depository for storage and withdrawal of measured amounts of energy needed for various life processes. It was early speculated that in photosynthesis, too, some light energy may be stored in this form. Some even suggested that formation of high energy phosphate may be the main, if not the only, mechanism of energy storage in photosynthesis. However, in this form the hypothesis is implausible. The high yield of energy utilization of photosynthesis seems incompatible with a mechanism in which the energy quanta provided by light (about 40 Kcal/einstein each), are first broken into "phosphate quanta" (about 8 Kcal/mole), to be reassembled later as "oxidation-reduction quanta" (about 120 Kcal/mole). However, high energy phosphate may well play an accessory role in photosynthesis. In reaction scheme 14.4, over 90% of the energy of photosynthesis is stored in the two "uphill" oxidation-reductions; but some ATP is formed as a side product in the intermediate enzymatic reaction chain, and used later to make certain endergonic follow-up reactions possible.

What is "high energy phosphate"? It is an ester of phosphoric acid that liberates 7–8 Kcal per mole upon hydrolysis, whereas the hydrolysis of "ordinary" or "low energy" phosphate esters is slightly endothermic.

Hydrolysis is the breaking of a chemical bond by water, with the water radicals (H) and (OH) attaching themselves to the open ends of the broken bond; examples are the hydrolysis of a C—C bond:

$$R_1R_2 + H_2O \rightarrow R_1H + R_2OH \qquad (18.1)$$

where R_1 and R_2 are two organic radicals; and hydrolysis of a C—O bond, such as

$$R_1\text{—O—}R_2 + H_2O \rightarrow R_1OH + R_2OH \qquad (18.2)$$

Hydrolyses do not change the reduction level (Chapter 5) of the hydrolyzed compounds, and, therefore, have only relatively small values of ΔH.

Hydrolysis of a phosphate ester is similar to reaction **18.2**, except that one (or both) organic radicals, R_1 and R_2, are replaced by univalent phosphate radicals, (P) = H_2PO_3, for example:

$$R_1\text{—O—(P)} + H_2O \rightarrow R_1OH + \text{(P)OH} \ (= H_3PO_4)^1 \qquad (18.3)$$

or

$$\text{(P)—O—(P)} + H_2O \rightarrow 2\text{(P)OH} \ (= 2H_3PO_4) \qquad (18.4)$$

The "high energy phosphates" differ from other phosphate esters because the products of their hydrolysis are stabilized by a kind of *resonance* that did not exist before hydrolysis. For example, in the case of a carboxylic acid, the hydrolysis of its phosphate is given by:

$$RC\begin{matrix}\nearrow\!\!\!\!\!\nearrow O \\ \searrow O\text{—(P)}\end{matrix} + H_2O \rightarrow RC\begin{matrix}\nearrow\!\!\!\!\!\nearrow O \\ \searrow OH\end{matrix} + \text{(P)—OH} \qquad (18.5)$$

This reaction leads to a carboxyl group, which is stabilized by resonance. This is easiest to illustrate on the example of its anion, which can exist in two "resonating" forms with the same energy content:

$$RC\begin{matrix}\nearrow\!\!\!\!\!\nearrow O \\ \searrow O^-\end{matrix} \quad \text{and} \quad RC\begin{matrix}\nearrow O^- \\ \searrow\!\!\!\!\!\searrow O\end{matrix}$$

differing only by distribution of electrons. It is a general rule that the existence of such "resonating" structures contributes to the stability of a molecule. Resonance stabilization of the carboxyl group makes carboxyl

[1] At other places in this book Pi is used, instead of P(OH), for H_3PO_4.

phosphates "high energy phosphates," compared to (for example) alcohol phosphates, $RCH_2O(P)$, because free alcohol formed by its hydrolysis is *not* stabilized by resonance:

$$RCH_2O(P) + H_2O \rightarrow RCH_2OH + (P)OH(=H_3PO_4)$$

Phosphate esters of the type (18.4) also are "high energy" phosphates, because free phosphoric acid is stabilized by a resonance similar to that present in carboxylic acids. For example, in the case of a univalent phosphate anion, $H_2PO_4^-$, the resonance is between the forms

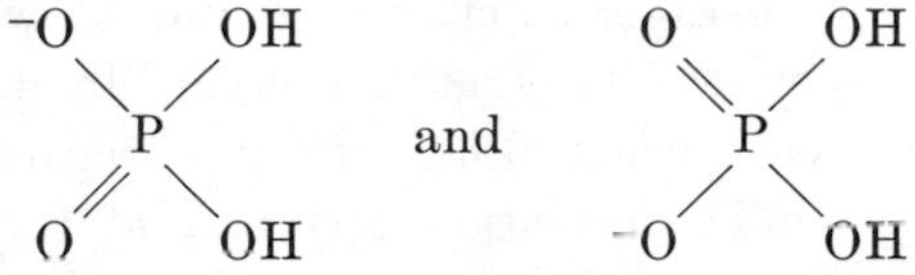

```
-O      OH               O      OH
  \    /                  \\    /
    P         and           P
  //   \                  /    \
O       OH             -O       OH
```

which differ only by electron distribution.

In metabolic processes, the most common high energy phosphate is *adenosine triphosphate* (ATP):

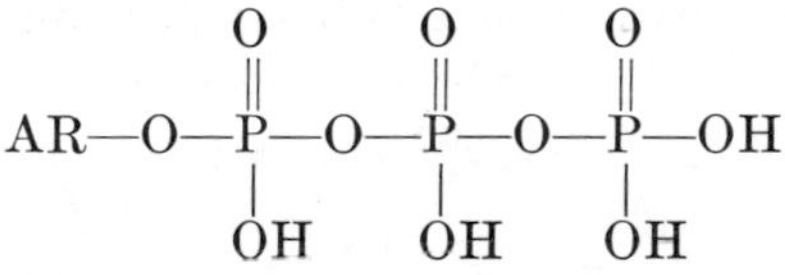

```
          O      O      O
          ||     ||     ||
AR—O—P—O—P—O—P—OH
          |      |      |
          OH     OH     OH
```

where AR is an organic base called adenosine, consisting of a nitrogeneous organic base, adenine, A (see below), to which a pentose sugar (ribose, R) is attached in the position marked by asterisk.

```
                NH2
                 |
                 C        N
               //  \    /   \\
             N       C        CH
Adenine:     |       ||       |
             HC      C——————NH*
               \\   /
                 N
```

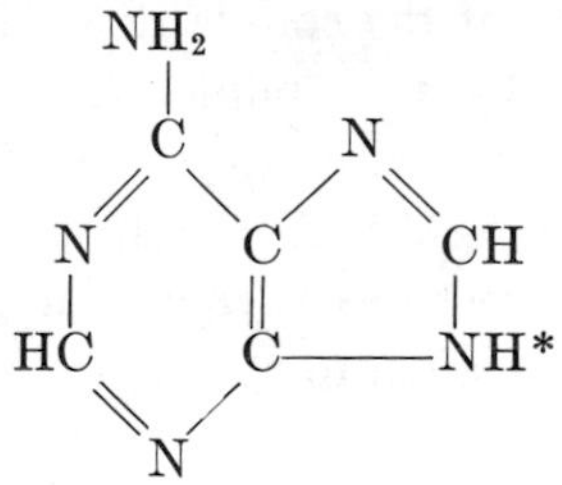

Why the principal biological energy carrier has to be attached to a specific organic base, why it includes a pentose sugar, and why it contains three phosphate radicals in a row, we don't know; the "high energy" character is not dependent on any of these characteristics. They may be important, however, from the point of view of enzymatic specificity, permitting convenient coupling of the conversion of ATP to ADP + (P) OH with an oxidation-reduction.

In mitochondria, a series of oxidation-reduction catalysts (including several cytochromes) are arranged in such a way that the transfer of an electron from one of them to the next one can be coupled with the formation of an ATP molecule from ADP and (P)OH. In this way, about one half of the free energy liberated in a whole series of oxidation-reductions is stored in ATP molecules. The latter diffuse away, to be hydrolyzed into ADP + (P)OH when and where energy is needed.

Recently, a theory became popular according to which the coupling of a redox reaction with ATP-synthesis occurs by the redox reaction establishing an H^+-ion concentration difference between two sides of a membrane; and the ATP formation occurring at the cost of the free energy of this gradient. (This is the so-called "chemiosmotic" theory of phosphorylation, proposed by the British chemist, P. Mitchell in 1961.) Experiments by André T. Jagendorf (then at Johns Hopkins University in Baltimore) showed that some ATP can, in fact, be formed by such a mechanism at the cost of an artificially established H^+ gradient. It remains, however, to be shown whether this mechanism of ATP formation can be as efficient as the ATP-synthesis actually is in respiration and in photosynthesis.

Earlier in the reaction sequence of respiration (before the cytochromes come into play) oxidation of phosphoglyceraldehyde (= triose phosphate), by nicotinamide adenine dinucleotide, to phosphoglyceric acid, PGA, also is coupled with conversion of ADP to ATP. The free energy available in the oxidation of the couple RCOOH/RCHO (with a potential of about—0.5 volt) by the couple NAD^+/NADH (with a potential of about —0.35 volt) is neatly stored in this coupled reaction. (The free energy available is 0.5–0.35 eV = 0.15 eV = 3.5 Kcal per electron, or 7.0 Kcal for the two electrons involved in oxidation of an aldehyde to an acid—which is just enough to synthesize one molecule of ATP.)

ATP-NEED IN PHOTOSYNTHESIS

The now widely accepted scheme of photosynthesis (Chapter 17) suggests that NADPH (reduced nicotinamide dinucleotide phosphate) and

ATP (adenosine triphosphate), supplied by the photochemical primary processes, produce a reversal of the reaction described at the end of the preceding section, that is, *reduce* phosphoglyceric acid (PGA) to phosphoglyceraldehyde (triose phosphate) and hydrolyze ATP to ADP and (P)OH. The "reducing energy" of NADPH and the hydrolysis energy of ATP together make the reaction 18.6 run in the desired direction.

$$\text{ATP} + \text{PGA} + \text{NADPH} \leftrightharpoons \text{Triose phosphate} + \text{NADP}^+ + \text{ADP} + \text{(P)OH} \qquad (18.6)$$

In the presence of the two enzymes that catalyze this reaction (triose dehydrogenase and kinase), reaction 18.6 will run from left to right, or from right to left, depending on relative concentrations of the reactants.

It was mentioned in Chapter 17 that another reaction, postulated in the Calvin cycle, also was found to require ATP—the introduction of a second phosphate group into ribulose monophosphate, to form ribulose diphosphate.

The Calvin cycle thus requires *three* ATP molecules to reduce *one* molecule of CO_2—one to form ribulose diphosphate, and thus to produce *two* molecules of PGA (see Eq. 17.4), and *two* to convert these two PGA molecules into two triose molecules. The Hatch—Slack pathway (see p. 239 in Chap. 17) requires an additional ATP.

Hydrogenation of *two* PGA molecules is needed because we must move *four* hydrogen atoms from H_2O to CO_2 to produce one O_2 molecule. Theoretically, this could be done in two steps—first, reducing a molecule of an acid, RCOOH, to an aldehyde, RCHO, and then reducing the latter to an alcohol, RCH_2OH. The second step requires considerably less energy than the first one (E_0' is —0.5 volt for the first step, and only —0.3 volt for the second one). However, according to Calvin's scheme, nature has chosen the hard way—that of reducing *two* RCOOH groups (in two PGA molecules). The excess energy stored in this way is regained, according to this scheme, in the dismutation associated with carboxylation, when ribulose diphosphate is converted to two PGA molecules (Eq. 17.4). In this way, the "waste" of light energy implied in the reduction of two carboxyls (instead of one carboxyl and one carbonyl), reveals itself as a clever trick. Ordinary carboxylations

$$\text{RH} + \text{CO}_2 \rightleftarrows \text{RCOOH} \qquad (18.8)$$

are endergonic and cannot run from left to right without supply of outside energy. Reaction (18.8), runs spontaneously from right to left (except under very high partial pressure of carbon dioxide, not available in nature). But reaction 17.4 is exergonic because dismutation is involved. By storing extra energy in two reduced carboxyl groups, nature found a way to utilize light energy indirectly for the carboxylation step.[2]

It was mentioned before that the primary acceptor of electrons in light reaction I may be an unknown compound (X), rather than the known intermediates, $NADP^+$, or ferredoxin. If PGA is reduced directly by XH (with an E_0'-value as low as —0.6 eV), the need for ATP can be eliminated. This may reduce the amount of ATP needed for the Calvin cycle to the one molecule needed for the phosphorylation of RMP to RDP.

Chloroplasts do produce ATP in light; this observation is to be discussed below. Whether the ATP yield obtainable in this way is high enough to permit the Calvin cycle to run in its original version (requiring 3 ATP per 4 electrons), or is sufficient only for the modified cycle (requiring 1 or 2 ATP's), remains an open question.

PHOTOSYNTHETIC PHOSPHORYLATION

Association of photo- and chemosynthesis with the conversion of adenosine diphosphate (ADP) and inorganic phosphate (P)OH into adenosine triphosphate (ATP) has been often suggested. Actual ATP formation from ADP and (P)OH was first observed in chemosynthetic bacteria: It was also noted in photosynthesizing cells by O. Kandler as uptake of (P)OH, and by B. Strehler (as ATP production revealed by firefly luminescence). Later (in 1954) it was found by D. Arnon, M. B. Allen, and F. R. Whatley (in Berkeley) in illuminated chloroplast prep-

[2] However, it may be that this is not at all the way it works. In some recent modifications, the Calvin cycle is supposed to involve the formation of only one PGA molecule, so that only one reaction reduces a carboxyl group in PGA, while the other may involve the reduction of a carbonyl; if both reductions use NADPH as reductant, ATP should be needed only in the "difficult" carboxyl reduction, and not in the "easy" carbonyl reduction; this would reduce the ATP requirement from 3 to 2. The above-suggested ingenious solution of the problem of "how to make carboxylation exergonic," will then have to be given up.

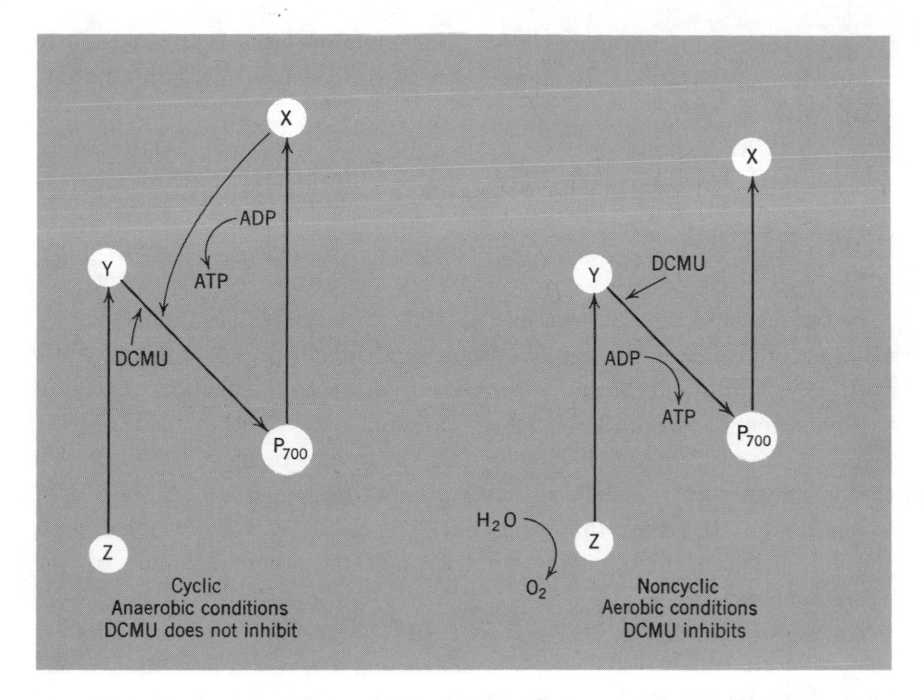

FIG. 18.1 Cyclic and noncyclic photophosphorylation. (Symbols have the same meaning as in Fig. 14.4; DCMU = 3-(3′,4′ dichlorophenyl)1,1 dimethyl urea.)

arations, and by A. Frenkel (in Minneapolis) in illuminated chromatophores from purple bacteria.

Arnon introduced a distinction between two kinds of photophosphorylation, which he called "cyclic" and "noncyclic." The first one is not coupled with on-going oxidation-reduction; the second one is (see Fig. 18.1).

In chloroplasts, cyclic phosphorylation occurs most efficiently when the Hill reaction is prevented; in whole cells, in the absence of photosynthesis.[3] Under these conditions, ATP formation is the only net photochemical change:

$$\text{ADP} + \text{(P)OH} \xrightarrow[\text{chlorophyll}]{\text{light}} \text{ATP} \qquad (18.9)$$

Well-known poisons of photosynthesis (such as 3-(3′,4′ dichlorophenyl) 1,1 dimethyl urea, DCMU), do not inhibit this kind of phosphorylation.

[3] Or when photosynthesis runs with less than the maximum yield. Cytochrome b_6 is an intermediate in this reaction.

"Noncyclic photophosphorylation" accompanies the Hill reaction in chloroplast preparations. It requires the presence of an electron acceptor (Hill oxidant), A.

$$\text{(P)OH} + H_2O + A + \text{ADP} \xrightarrow[\text{chlorophyll}]{\text{light}} O_2 + AH_2 + \text{ATP} \quad (18.10)$$

DCMU inhibits this type of phosphorylation (by 50% in concentrations as low as 2×10^{-7}M).

In the mechanism suggested in Fig. 18.1, "noncyclic" photophosphorylation is coupled with the main reaction sequence of photosynthesis, while cyclic photophosphorylation is associated with a back reaction, reversing the hydrogen transfer in PS I. (We recall that photosynthesis is a process "against nature," in which H-atoms or electrons are moved "uphill," so that a danger of "falling back" and dissipating the stored energy lurks at each step. It seems that such back reactions can be coupled with the formation of high energy phosphate, thus salvaging some of the wasted energy.)

An experiment by A. T. Jagendorf and co-workers (now at Cornell) suggested that light-induced phosphorylation occurs in two steps: one photochemical, the other dark. They preilluminated chloroplasts and transferred them quickly (within seconds) into another vessel (kept in darkness), containing ADP, (P)OH and Mg^{++}. The synthesis of ATP now took place. Light must have produced some long-lived high energy intermediate, capable of synthesizing ATP from ADP and P(OH) in darkness.

As shown in Fig. 18.1, at least one molecule of ATP can be formed as a "free premium" in the passage of two electrons through the main oxidation-reduction chain, without requiring a diversion of quanta. Since the main process uses 8 quanta to send four H-atoms from H_2O to CO_2, only two ATP molecules can be obtained in this way per CO_2 reduced. This is not enough to make the Calvin cycle (or the Hatch and Slack pathway; see p. 239) run in its original version, requiring 3 (or 4) ATP's per CO_2 reduced; but just enough to run it in the amended way (as discussed above).

If one postulates that ATP is formed in photosynthesis by the "chemiosmotic" mechanism, utilizing the energy of several electron transfers (each liberating less energy than is needed to form one ATP), then the enzymatic redox reaction sequence that follows the photochemical step [X → Fd → NADP] also could be utilized for this purpose, even if it liberates only 0.25 eV per electron.

Kok has suggested from kinetic evidence that the oxidant of PSII has a normal potential of +0.18 volt, so that only 0.4–0.18 = 0.22 eV (rather than 0.4 eV) become available in the intermediate enzymatic sequence between PSI and PSII. Measurements by W. A. Cramer and W. Butler suggested, however, that the correct value is —0.03 volt; they also found evidence that another intermediate, with a potential of —0.27 volt, may exist, the available energy is thus 0.44 or even 0.67 eV.

Experimentally, quantum requirements of 6 have been found per ATP-molecule formed, for "cyclic" and "noncyclic" photophosphorylation alike. If photophosphorylation were to occur in a separate photochemical reaction, this would mean the need of 12 (or 18) quanta just to supply the two (or three) ATP molecules needed for the Calvin cycle. Even with two ATP's contributed as a side product when 8 quanta are used for oxidation-reduction, this mechanism would still require 6 additional quanta to produce the ATP molecules needed for the original Calvin cycle—a requirement incompatible with the observed overall quantum requirement of 8! (Answers to these questions may be found only when simultaneous, precise quantum yield measurements for ATP production and oxygen evolution will be made.)

These calculations emphasize the uncertainty of the widely accepted assumption that the photochemical process in photosynthesis supplies both the NADPH and all the ATP needed for the Calvin cycle. This assumption is based on two sets of facts: (a) chloroplasts *do* contain sufficient amounts of $NADP^+$, and *can* convert it in light to NADPH, and *do* transform ADP to ATP; and (b) enzymatic mechanisms *can* convert carbon dioxide to carbohydrates, if enough NADPH and ATP is supplied. However, to be certain that a model mechanism suggested for photosynthesis is truly adequate, one has to prove that it can run, under natural conditions, with the high efficiency characteristic of natural photosynthesis—and this proof is as yet missing for the ATP-requiring steps in the Calvin cycle (at least, in its original version).

Chapter 19

Summary and Outlook

We have arrived at the end of our story. It is a highly provisional end; but then, the picture only started to unfold itself three decades ago. Until then, only two fundamental facts were known about the mechanism of photosynthesis: (1) that it consists of a photochemical stage and an enzymatic "dark" stage, with the latter imposing a "ceiling"—a maximum rate at which the overall reaction can proceed in strong light (F. F. Blackman, 1905); and (2) that the dark reaction requires about 10^{-2} sec for its completion (at 20°C), and has a maximum yield (as measured after a strong, "saturating," light flash) corresponding to about one molecule of O_2 produced by a single flash per 2500 molecules of chlorophyll present. These observations (Robert Emerson and W. Arnold, 1932) were interpreted (Hans Gaffron and K. Wohl, 1936) as indicating the existence of "photosynthetic units" of $2500/n$ chlorophyll molecules (n = a small number—perhaps 4 or 8), containing a single enzymatic "reaction center."

The first spurt in the recent progress of understanding of the mechanism of photosynthesis followed the discovery of the long-lived radioactive carbon isotope, ^{14}C, by Sam Ruben and Martin Kamen in 1940. In the hands of Melvin Calvin and Andrew Benson, ^{14}C became a master tool for unraveling the complex enzymatic mechanism of carbon dioxide transformation into carbohydrate. This development still goes on, but the two main facts—the pivotal roles of PGA (phosphoglyceric acid) and RDP (ribulose disphosphate), were established already in 1949 and 1952, respectively.

Another flood of new knowledge has begun to spread concerning what is, perhaps, the most exciting aspect of photosynthesis—its primary photophysical and photochemical process. From the empirical picture of a "photosynthetic unit," there developed a theoretical concept of *excitation energy migration*—either among identical molecules, by a "first order" (or "exciton") resonance mechanism, or between different molecules, by a "second order" ("Förster") resonance mechanism. This migration makes it possible for quanta absorbed by any pigment molecule within a unit to initiate a sequence of enzymatic reactions in a single common "reaction center."

Photosynthesis now appears as a tripartite reaction. One part of it is the *evolution of oxygen from water,* leaving behind an unknown reduced compound ZH (Fig. 5.4). The second is the *"uphill" transfer of H-atoms* (or electrons) from ZH to the—also not definitely known—primary hydrogen acceptor, X. The third part is *reduction of* CO_2 by XH, leading to hexose and starch. (This third part is the one elucidated by the carbon tracer experiments.)

The light quanta captured by the pigments are used to perform the second, energy-storing step. In other words, light energy is utilized in photosynthesis for an *oxidation-reduction reaction against the gradient of chemical potential.*

We now have several reasons to believe that the photochemical, energy-storing stage in photosynthesis itself consists of *two separate photochemical steps,* I and II, connected by an enzymatic reaction sequence; the two are sensitized by two separate pigment systems, "PSI" and "PSII."

Most convincing in this connection are two sets of observations. The first is Robert Emerson's measurements (with Charlton Lewis) of the *quantum requirement of photosynthesis* (1938–42); they suggested that a minimum of *eight* quanta is needed to transfer *four* hydrogen atoms (or electrons) from H_2O to CO_2, reducing the latter to the carbohydrate level, and liberating one molecule of oxygen. His observation, in 1943, of the *red drop* of this efficiency at the long-wave end of the visible spectrum, and discovery (with R. Chalmers and C. Cederstrand in 1957) that the rate of oxygen evolution in the region of the "red drop" is enhanced when illumination with far-red light is supplemented by a beam of higher frequency light, led to the conclusion that two consecutive four-quanta processes (two consecutive uphill transfers of four electrons or H-atoms), are involved in photosynthesis, and that they are brought about by two "pigment systems," containing a slightly different assortment of pigments (Fig. 13.9). The two systems must operate at the same rate for photosynthesis to run with maximum efficiency; the "red drop" occurs when absorption in one pigment system (PSII) becomes much weaker than that in the other (PSI). It disappears when additional illumination improves the absorption in PSII (*enhancement effect*).

A second argument for the two-step concept of the primary photochemical process was derived by Robert Hill and F. Bendall in 1960 from consideration of the oxidation-reduction potentials of the cytochromes (Fig. 14.4). The finding (by L. N. M. Duysens, J. Amesz, and B. M. Kamp in 1961) that light absorbed by one pigment system (PSII—specifically, by the phycobilins in red algae) causes the reduction, and light absorbed by another pigment system (PSI—specifically, by chlorophyll *a* in red algae) causes the oxidation of certain cytochromes provided one of the best pieces of evidence for two light reactions in photosynthesis. (Cytochromes, long known to play a key role in respiration, were found in 1952, by R. Hill and co-workers, to be present in all photosynthesizing cells.) Hill and Bendall suggested that one pigment system (PSII) sensitizes the oxidation of water (normal redox potential, +0.8 volt) by a reduced cytochrome (normal potential, about 0.0 volt), and the other (PSI), the oxidation of a reduced cytochrome (with a normal potential of about +0.4 volt) by an organic oxidant with a normal potential of about —0.4 volt, such as oxidized pyridine nucleotide (and indirectly, by carbon dioxide).

A great amount of work was devoted recently to the enzymatic reaction sequence connecting the two photochemical processes—a "downhill" electron transfer from $E_0' = 0.0$ volt to $E_0' = +0.4$ volt. Several components apparently involved in this reaction sequence have been identified. Among these are plastoquinone and plastocyanin, in addition to two cytochromes, cytochrome *f* and (probably) cytochrome b_3; and an as yet mysterious "compound *Q*." Difference spectroscopy has led to tentative identification of the "energy traps" (reaction centers) postulated in the two pigment systems (P700 in PSI identified by Bessel Kok in 1956, and P690 in PSII, identified in H. T. Witt's laboratory in 1967).

From electron microscopy, the conclusion was drawn that photosynthetic units probably are real structural elements of the chloroplasts (rather than, what was a priori a possibility, merely statistical ratios between the numbers of pigment molecules and those of enzymatic reaction centers in them). They are now thought to consist of about 200–300 chlorophyll molecules. Each unit includes accessory pigments (such as chlorophyll *b*, carotenoids, or phycobilins), several varieties of chlorophyll *a*, and a single reaction center ("energy trap"). Light energy absorbed by a pigment molecule anywhere in the photosynthetic unit is transferred (by the above-mentioned resonance mechanisms) to this trap, where it can be utilized for the energy-storing primary photochemical reaction. Units of one type (PSII), which bring about the oxidation of ZH and reduction of a cytochrome, contain most of the accessory pigments (including chlorophyll *b* in green plants, and phycobilins in red and blue-green algae), slightly more of a "short-wave form" of chlorophyll *a* (Chl *a* 670), and slightly less of "long-wave" form, Chl *a* 680 than units of type PSI, and a "trap," P690; while units of the second type (PSI), which bring about the oxidation of a cytochrome and the reduction of X, contain much less accessory pigments, somewhat less Chl *a* 670, and somewhat more of the long-wave form of Chl *a* (Chl *a* 680). They also contain a Chl *a* form absorbing at still longer waves (Chl *a* 695), and a "trap," P700.[1]

In light reaction II, a strong oxidant (Z) is produced, that can oxidize H_2O to O_2, and an intermediate reductant (such as a reduced cytochrome b_3); in light reaction I, a strong reductant is produced that can reduce CO_2 to carbohydrate, and an intermediate oxidant (such as oxidized cytochrome *f*). Formation of high energy phosphate, ATP, is coupled to the reaction connecting light stages II and I—reaction between ferricytochrome *f* in PSI and the ferrocytochrome b_3 in PSII. This reaction seems to involve, in addition to the two cytochromes, several other intermediate catalysts. A back reaction in PSI also seems able to produce ATP; but this can occur only when photosynthesis runs at less than full maximum quantum efficiency, because it requires a part of the products of reaction I to be lost for the overall oxidation-reduction process.

Two products of the light reactions—high energy phosphate, ATP, and a strong reductant (such as NADPH)—act together to reduce CO_2

[1] In red and blue-green algae, Chl*a*670 and Chl*a*680, are largely replaced, in PSII, by phycobilins.

to the carbohydrate (CH_2O) level. In this reaction sequence, as elucidated by Calvin and co-workers, CO_2 is first added to a five-carbon sugar phosphate (ribulose diphosphate), thus producing two molecules of phosphoglyceric acid (PGA); these are then reduced to two molecules of a three-carbon sugar (triose) by NADPH (or another equally potent, or even more potent, photochemically generated reductant), with the help of the hydrolysis energy of ATP. Two molecules of the so-formed triose are transformed into one molecule of a six-carbon sugar (fructose or glucose). The CO_2-"carrier," ribulose diphosphate, needed for the next cycle, is regenerated from additional molecules of the synthesized triose, by another series of enzymatic reactions (Calvin-Benson cycle).

There is no certainty that the enzymatic mechanism of CO_2-reduction to sugar, developed by Calvin and co-workers, is unique. Rather, as in the case of the reverse process, that of cellular respiration, several alternative enzymatic pathways may exist. One such mechanism, involving phosphoenolpyruvic acid as primary carbon dioxide absorber (instead of ribulose phosphate) was proposed recently (1966) by M. D. Hatch and C. R. Slack for corn, sugar cane, and some other higher plants. Such alternative pathways may account for occasional appearance of compounds not in the Calvin-Benson cycle, such as malic acid, among early heavy-labeled products in C*-tracer experiments.

One may ask whether the suggested two-step mechanism of the primary photochemical process also allows variations. Undoubtedly, under certain conditions, only one of the two systems is operative, while the other idles, or is inhibited (or, in the case of exclusive excitation of one pigment system, is not involved at all). One or both stages may take part in bacterial photosynthesis, and in "artificial" variants of the natural process, such as the Hill reaction.

Even leaving all possible variants aside, the above-described tripartite mechanism of photosynthesis is not yet established beyond all challenge; only future experiments can decide which parts of it will survive and which will have to be modified or abandoned.[2] Nevertheless, compared to the total ignorance that had prevailed before 1940 (when the photosynthetic cell appeared as a magic black box, taking up CO_2, H_2O,

[2] Alternative formulation include schemes in which one light reaction makes ATP, and the other NADPH, and schemes in which both light reactions produce oxidants and reductants, but the two oxidants somehow cooperate to oxidize H_2O and the two reductants cooperate to reduce CO_2.

and light and pouring out sugar and oxygen), the development of this mechanism represents an impressive advance achieved in the last quarter of a century.

Important questions that will have to be tackled in the future include the detailed physical mechanism of energy migration in the photosynthetic units, the physicochemical mechanism of the primary photochemical steps, and the details and alternatives of the enzymatic mechanism. The latter include, in particular, the coupling of ATP synthesis and hydrolysis with oxidation-reductions, the function of cytochromes and quinones in the intermediate enzymatic reaction sequence, and the—all but unknown—mechanism of oxygen evolution. The locations of the components of the two systems in the chloroplast structure (which we now know to be lamellar) remain to be established; so is the whole "topochemistry" of the three stages of the overall process. Promising in that connection are experiments by Boardman and Anderson (1966) suggesting the possibility *of mechanical separation* of PSI and PSII or at least, their substantial enrichment in certain fractions of chloroplast material.

The ultimate possibility of *in vitro photosynthesis* is beckoning. Ideally, this should take the form of combined storage of light energy *and* synthesis of organic matter from inorganic materials; more modestly, but still very importantly, one can aim at finding a way to store efficiently light energy as utilizable chemical energy in a relatively simple (inorganic or organic) chemical system.

The *evolutionary history* of photosynthesis remains a fascinating question. Has chemosynthesis preceded it on earth? Did bacterial photosynthesis, utilizing relatively unstable "energy-rich" hydrogen sources, such as H_2S or H_2, in turn precede "true" photosynthesis (which makes use of the most abundant, but also the most reluctant hydrogen donor on earth, water)? This hypothesis fits the widely accepted Oparin-Haldane theory of the origin of life on earth, which postulates the early existence on earth of a reducing atmosphere (an atmosphere containing H_2, H_2S, and hydrocarbons, but no O_2). Plant metabolism is supposed to have first enriched the air with oxygen. However, bacterial photosynthesis (and bacterial chemosynthesis), as they occur now, may well represent not the survival of early forms of life, but relatively recent products of adaptation of early photosynthetic organisms to a "lazy" way of life in energy-rich, reducing media.

Could life have developed on other cosmic bodies in a different way from that which it had followed on earth? We have every reason to assume that the same chemical elements are present throughout the universe. Can one imagine life based on elements other than carbon, oxygen, and hydrogen? Could, for example, silicon substitute for carbon, to make biological materials viable at much higher temperatures outside the "biological range" of temperatures on earth? Could ammonia function, at low temperatures, as a substitute of water? Such speculations are tempting, but unlikely. The capacity of silicon to imitate carbon in the formation of complex molecules is quite limited, even if in recent years, the chemistry of silicones has become an important branch of industrial chemistry.

If life *must* be based on the elements C, H, and O, it can only exist between 0 and 100°C—that means, only on moderately warm cosmic bodies. Also, it can only develop on planets not much lighter than the earth, since otherwise the orignally available hydrogen would have dissipated in space soon after the birth of the planet.

Altogether, there are many fascinating problems left related to, or arising from, the chemical and physical mechanisms of photosynthesis, its evolution, its control, and its imitation outside the living cell.

Bibliography

We give below a list of books, collections of papers, Symposia, Reviews, and Scientific American articles. This is by no means an exhaustive bibliography; but references to original papers can be found in these publications.

Beginning students may first read the Scientific American articles, and advanced students, the latest reviews.

BOOKS

Year

1945 1. E. Rabinowitch: *Photosynthesis and Related Processes,* Vol. I, 599 pp., Wiley (Interscience), New York, 1945.

1951 2. E. Rabinowitch: *Photosynthesis and Related Processes,* Vol. II, Part I, 605 pp., Wiley (Interscience), New York, 1951.

1955 3. R. Hill and C. P. Whittingham: *Photosynthesis,* 164 pp., Methuen and Co., Ltd., London, 1955.

1956 4. E. Rabinowitch: *Photosynthesis and Related Processes,* Vol. II, Part 2, 879 pp., Wiley (Interscience), New York, 1956.

1957 5. J. A. Bassham and M. Calvin: *The Path of Carbon in Photosynthesis,* 104 pp., Prentice-Hall, Englewood Cliffs, New Jersey, 1957.

1958 6. E. Rabinowitch: *"La photosynthèse,"* 172 pp., Gauthier-Villars, Paris, 1958.

1960 7. W. Ruhland (ed.): *"The Assimilation of Carbon Dioxide,"* in *Encyclopedia of Plant Physiology,* Vol. V, Part 1, 1013 pp., Part 2, 868 pp., Springer-Verlag, Berlin, 1960.

1960 8. H. Gaffron: *Energy Storage: Photosynthesis* in *Plant Physiology* (F. C. Steward, ed.), Vol. IB (pp. 3–277), Academic Press, New York, 1960.

Year

1962 9. M. Calvin and J. A. Bassham: *The Photosynthesis of Carbon Compounds,* 127 pp., W. A. Benjamin, Inc., New York, 1962.

1963 10. M. D. Kamen: *Primary Processes in Photosynthesis,* 183 pp., Academic Press, New York, 1963.

1963 11. Japanese Society of Plant Physiologists (ed.): *Studies on Microalgae and Photosynthetic Bacteria,* 636 pp., The University of Tokyo Press, Tokyo, 1963.

1964 12. A. C. Giese (ed.): *Photophysiology,* Vol. I, 377 pp., Academic Press, New York, 1964.

1965 13. R. K. Clayton: *Molecular Physics in Photosynthesis,* 205 pp., Blaisdell Publishing Co., New York, 1965.

1965 14. J. B. Thomas: *Primary Photoprocesses in Biology,* 323 pp., North Holland Publishing Co., Amsterdam, 1965.

1965 15. J. L. Rosenberg: *Photosynthesis, the Basic Process of Food-making in Green Plants,* 127 pp., Holt, Rinehart and Winston, New York, 1965.

1966 16. L. P. Vernon and G. R. Seely (eds.): *The Chlorophylls—Physical, Chemical, and Biological Properties,* 679 pp., Academic Press, New York, 1966.

1966 17. D. R. Sanadi (ed.): *Current Topics in Bioenergetics,* Vol. 1, 292 pp., Vol. 2, 373 pp., Academic Press, New York, 1966–1967.

1968 18. G. E. Fogg: *Photosynthesis,* 116 pp., American Elsevier Publishing Co., New York, 1968.

SYMPOSIA

1935 1. Cold Spring Harbor Symposia on Quantitative Biology, Vol. III, 359 pp., The Biological Laboratory, Cold Spring Harbor, New York, 1935.

1949 2. J. Franck and W. E. Loomis (eds.): *Photosynthesis in Plants,* 500 pp., Iowa State College Press, Ames, Iowa, 1949.

1951 3. J. F. Danielli and R. Brown (eds.): *Carbon Dioxide Fixation and Photosynthesis,* Symposia for Experimental Biology, No. 5, 342 pp., The University Press, Cambridge, 1951.

1957 4. H. Gaffron, A. H. Brown, C. S. French, R. Livingston, E. I. Rabinowitch, B. L. Strehler, and N. E. Tolbert (eds.): *Research in Photosynthesis* (Symposium held in 1955), 524 pp., Interscience, New York, 1957.

1958 5. R. C. Fuller (Chairman): *The Photochemical Apparatus—Its Structure and Function,* Brookhaven Symposia in Biology, No. 11, 366 pp., Brookhaven National Laboratory, Upton, New York, 1958.

Year

1959 6. Biological Section of U.S.S.R. Academy of Science and Biology Department of Moscow State University: *Problems of Photosynthesis* (Symposium held in 1957), 747 pp., Publishing House of U.S.S.R. Academy of Sciences, Moscow, 1959.

1960 7. M. B. Allen (ed.): *Comparative Biochemistry of Photoreactive Systems,* Symposia on Comparative Biology, Vol. 1, 437 pp., Academic Press, New York, 1960.

1960 8. L. J. Heidt, R. S. Livingston, E. Rabinowitch, and F. Daniels (eds.): *Photochemistry in the Liquid and Solid States* (Symposium held in 1957), 174 pp., John Wiley and Sons, New York, 1960.

1961 9. W. D. McElroy and B. Glass (eds.): *A Symposium on Light and Life* (held in 1960), 924 pp., Johns Hopkins Press, Baltimore, 1961.

1961 10. B. Chr. Christensen and B. Buchmann (eds.): *Progress in Photobiology—Proceedings of the 3rd International Congress on Photobiology* (held in 1960), 628 pp., Elsevier Publishing Co., Amsterdam, 1961.

1963 11. H. Tamiya (ed.): *Mechanism of Photosynthesis—Proceedings of the V International Congress of Biochemistry* (held in 1961), Symposium Vol. VI, 385 pp., PWN—Polish Scientific Publishers, Warsaw and Pergamon Press, London, 1963.

1963 12. H. Gest, A. San Pietro, and L. P. Vernon (eds.): *Bacterial Photosynthesis,* a symposium sponsored by the Charles F. Kettering Research Laboratory, 523 pp., The Antioch Press, Yellow Springs, Ohio, 1963.

1963 13. M. R. Wurmser (Chairman): "La photosynthèse," (Symposium held in 1962), Colloques Internationaux du Centre National de la Recherche Scientifique, No. 119, 645 pp., C.N.R.S., 15 Quai Anatole-France, Paris, 1963.

1963 14. B. Kok and A. T. Jagendorf (eds.): *Photosynthetic Mechanisms of Green Plants* (Symposium held in 1963), National Academy of Sciences—National Research Council Publication No. 1145, 766 pp., Washington, D.C., 1963.

1965 15. D. W. Krogmann and W. H. Powers (eds.): *Biochemical Dimensions of Photosynthesis,* 124 pp., Wayne State University, Detroit, 1965.

1965 16. E. J. Bowen (ed.): *Recent Progress in Photobiology* (Symposium held in 1964), 400 pp., Academic Press, New York, 1965 (also Blackwell Scientific Publishers, Oxford).

1966 17. T. W. Goodwin (ed.): *Biochemistry of Chloroplasts—Proceedings of a NATO Advanced Study Institute,* held in 1965, Vol. 1, 476 pp., Vol. 2, 700 pp., Academic Press, New York, 1966–1967.

1966 18. J. B. Thomas and J. H. C. Goedheer (eds.): *Currents in Photosynthesis—Proceedings of the 2nd Western Europe Conference on Photo-*

Year

synthesis, held in 1965, 487 pp., Ad Donker Publishers, Rotterdam, 1966.

1967 19. J. M. Olson (Chairman): *Energy Conversion in Photosynthesis* (Symposium held in 1966), Brookhaven Symposia on Biology, No. 19, 514 pp., Brookhaven National Laboratory, Upton, New York, 1967.

1967 20. A. San Pietro, F. A. Greer, and T. J. Army (eds.): *Harvesting the Sun—Photosynthesis in Plant Life,* 342 pp., Academic Press, New York, 1967.

1967 21. K. Shibata, A. Takamiya, A. T. Jagendorf, and R. C. Fuller (eds.): *Comparative Biochemistry and Biophysics of Photosynthesis,* 460 pp., University Park Press, Baltimore, 1968.

1969 22. H. Metzner (ed.): *Progress in Photosynthesis Research—Proceedings of the 1st International Congress on Photosynthesis Research* (Symposium held 4–10 June, 1968, at Freudenstadt, W. Germany), 3 volumes, in press.

REVIEWS

1958 1. R. Emerson: "The Quantum Yield of Photosynthesis, "*Ann. Rev. Plant Physiol.,* **9** (1958), 1–24.

1961 2. G. Hoch and B. Kok: "Photosynthesis," *Ann. Rev. Plant Physiol.,* **12** (1961), 155–194.

1962 3. C. B. van Niel: "The Present Status of the Comparative Study of Photosynthesis," *Ann. Rev. Plant Physiol.,* **13** (1962), 1–26.

1962 4. A. T. Jagendorf: "Biochemistry of Energy Transformation During Photosynthesis," *Survey of Biological Progress,* **4** (1962), 181–344.

1963 5. J. H. C. Smith and C. S. French: "The Major and Accessory Pigments in Photosynthesis," *Ann. Rev. Plant Physiol.,* **14** (1963), 181–224.

1963 6. R. K. Clayton: "Photosynthesis: Primary Physical and Chemical Processes," *Ann. Rev. Plant Physiol.,* **14** (1963), 159–180.

1964 7. J. Franck and J. L. Rosenberg: "A Theory of Light Utilization in Plant Photosynthesis," *J. Theoretical Biology,* **7** (1964), 276–301.

1964 8. L. P. Vernon: "Bacterial Photosynthesis," *Ann. Rev. Plant Physiol.,* **15** (1964), 73–100.

1964 9. G. W. Robinson: "Quantum Process in Photosynthesis," *Ann. Rev. Physical. Chem.,* **15** (1964), 311–348.

1964 10. L. N. M. Duysens: "Photosynthesis," *Progress in Biophysics,* **14** (1964), 3–104.

Year

1965 11. A. San Pietro and C. C. Black: "Enzymology of Energy Conversion in Photosynthesis," *Ann. Rev. Plant Physiol.,* **16** (1965), 155–174.

1965 12. Govindjee and E. Rabinowitch: "The Photochemical Stage of Photosynthesis," *J. Scientific and Industrial Research* (New Delhi), **24** (1965), 591–596.

1965 13. L. N. M. Duysens: "On the Structure and Function of the Primary Reaction Centers of Photosynthesis," *Arch. Biol.* (Liège), **76** (1965), 251–275.

1965 14. R. K. Clayton: "Biophysical Aspects of Photosynthesis," *Science,* **149** (1965), 1346–1354.

1965 15. L. P. Vernon and M. Avron: "Photosynthesis," *Ann. Rev. Biochem.,* **34** (1965), 269–296.

1966 16. N. I. Bishop: "Partial Reactions of Photosynthesis and Photoreduction," *Ann. Rev. Plant Physiol.,* **17** (1966), 185–208.

1967 17. L. N. M. Duysens and J. Amesz: "Photosynthesis" in *Comparative Biochemistry,* M. Florkin and E. H. Stotz (eds.), **27** (1967), 237–266.

1967 18. Govindjee, G. Papageorgiou, and E. Rabinowitch: "Chlorophyll Fluorescence and Photosynthesis," in *Fluorescence, Theory, Instrumentation, and Practice,* Marcel Dekker, Inc., New York, 1967, 511–564.

1967 19. Govindjee: "Transformation of Light Energy into Chemical Energy; Photochemical Aspects of Photosynthesis," *Crop Science,* **7** (1967), 551–560.

1967 20. M. Gibbs: "Photosynthesis," *Ann. Rev. Biochem.,* **36II** (1967), 757 784.

1968 21. M. Avron and J. Neumann: "Photophosphorylation in Chloroplasts," *Ann. Rev. Plant Physiol.,* **19** (1968), 137–166.

1968 22. D. S. Bendall and R. Hill: "Haem-Proteins in Photosynthesis," *Ann. Rev. Plant Physiol.* **19** (1968), 167–186.

1968 23. G. Hind and J. M. Olson: "Electron Transport Pathways in Photosynthesis," *Ann. Rev. Plant Physiol.,* **19** (1968), 249–282.

1968 24. N. K. Boardman: "The Photochemical Systems of Photosynthesis," in *Advances in Enzymology,* F. F. Nord (ed.), 1968, 1–79.

ARTICLES IN "SCIENTIFIC AMERICAN"

1948 1. E. Rabinowitch: "Photosynthesis," *Sci. Am.,* **179** (1948), 3–13.

1953 2. E. Rabinowitch: "Progress in Photosynthesis," *Sci. Am.,* **189** (1953). 80–84.

Year

1960 3. D. I. Arnon: "The Role of Light in Photosynthesis," *Sci. Am.*, **203** (1960), 104–118.

1962 4. J. A. Bassham: "The Path of Carbon in Photosynthesis," *Sci. Am.*, **206** (1962), 88–100.

1965 5. E. I. Rabinowitch and Govindjee: "The Role of Chlorophyll in Photosynthesis," *Sci. Am.*, **213** (1965), 74–83.

Index